I0045043

Kishodai Tokyo Kanku

Report on the Meteorological Observations Made at High Level Stations in Japan

Kishodai Tokyo Kanku

Report on the Meteorological Observations Made at High Level Stations in Japan

ISBN/EAN: 9783744790864

Printed in Europe, USA, Canada, Australia, Japan

Cover: Foto ©berggeist007 / pixelio.de

More available books at **www.hansebooks.com**

REPORT

ON THE

METEOROLOGICAL OBSERVATIONS

MADE AT

HIGH LEVEL STATIONS

IN

JAPAN.

CENTRAL METEOROLOGICAL OBSERVATORY
OF JAPAN,
TOKIO.
1893.

CONTENTS.

REPORT

ON THE

METEOROLOGICAL OBSERVATIONS MADE AT HIGH LEVEL STATIONS

IN

JAPAN.

INTRODUCTION.

Amongst several important questions in the field of modern meteorology, the study of upper air stratum deserves of course to be reckoned, so that the Roman Congress in 1879, approving Dr. Hann's proposal, has recommended the establishment of high level stations in each country and the temporary observations to be made in the balloon ascents or during the expedition on the high mountains.

According to this decision some important stations have been erected in Europe and America, of which the principal are the following:

Name of Locality:	Altitude:
Mont Blanc	4,810 metres
Pike's Peak	4,340
Sonnblick	3,095
Pic du Midi	2,877
Säntis	2,690
Massachusetts	2,550
Val Dobbia	2,548
Saint-Bernard	2,478
Mont Koïlamsk	2,364
Darjeeling	2,107
Saint-Gothard	2,093
Obir	2,041
Simplon	2,008
Mount Washington	1,938
Mont Ventoux	1,906
Mont Aigoual	1,571
Puy de Dôme	1,463
Ben Nevis	1,460

Yet in the Eastern Asia, there is a great chasm in the view of this point, particularly in our country where the most complete meteorological *réseau* has been organized during past 10 years. We have now fifty stations, distributed between 26° and 46° of north latitude and between 128° and 146° of east longitude. Their altitudes vary as follows:

Altitude:	Number of Stations:
420—200 metres	1
200—100	3
100— 50	3
50 — 30	10
30— 2	33

Our country being mountainous in the interior, we have many favorable peaks suitable for this purpose, heights exceeding 3,000 metres at some prominent summits; as for example, Mount Fuji, the highest mountain (3,700 metres) in our country, affords much advantages according to its accessibility as well as its situation near the southern coast.

In spite of a great endeavor availed by our meteorologists, the scheme of the erection of high level stations has met always some difficulties; therefore we gratify for the moment in sending only during the hot season the expedition on the high mountains for the purpose of making simultaneous observations with the lower stations. But on account of uneasiness, the members of these expeditions could not stay no more longer than five or six weeks; consequently the data referred to these investigations are limited to the period extending from the beginning of August to the end of September.

In publishing this report we would not venture to deduce prematurely any conclusions from the results furnished by the members of the expeditions, but we have reserved only to collect in one pamphlet all the documents concerning the high level observations made by the members of the Observatory and formerly published in the separate papers. Besides these, we have also added some observations made individually by some meteorologists of our country on the high mountains. Therefore this reports contains at first the *résumés* of the comparative observations at high mountains and at corresponding lower stations, in the appendix the original documents obtained at the upper and lower stations.

Although this report comprises no specified consequences, yet we think it gives some interests to those whom the studies of meteorological elements in the high level are to be specially occupied. The preparation of this report relies almost, in its entirety, upon the reports on the Fuji, Ontake, Gozaishodake and Higashi-Hōben observations made by Messrs. Nakamura, Masato, Shimono, Kondo of our Observatory and Mr. Tonno of Yamaguchi Higher Middle School, as well as the reports of lower stations who participated to these investigations in making supernumerary observations during the expeditions. We express here our profound gratitude to all the members of the expeditions, to Mr. Tonno and to the observers of the meteorological stations who have made these interesting observations. For the preparation of this report Mr. Wada, Chief of the Service of Indications, has taken a greater part, to whom we express our best thanks for this valuable work.

K. KOBAYASHI,
The Director

January, 1893.

I. HISTORY OF HIGH LEVEL OBSERVATIONS IN JAPAN.

Besides the individual observations of some meteorological elements made by the foreign medical men in the Settlements of Yokohama, Nagasaki and Hakodate from an earlier time, the first regular observations were commenced in 1872; in this year the Government engaged some American agriculturists for the clearings of the Island of Hokkaido (namely Yezo), and by their request the Government has established the first meteorological station at Hakodate on the southern coast of this Island. Thereafter, the necessity of meteorological investigations being appreciated by the Government, the number of stations has increased gradually and in the period of ten years, that is in the end of the year 1882, it reached more than twenty; now the number of stations became just fifty.

But the high level observations were not commenced before the year 1887, excepting the barometric observations made by some Fuji accensionists for a single determination of its altitude. In this year Messrs. Knipping and Masato of our Observatory were sent on the summit of Fuji, accompanying Dr. Todd of Amherst Observatory, who visited Japan for the observations of Total Solar Eclipsis of August 19th. This was the first expedition of the official character, but having been interrupted by some special business, only three days were devoted for the meteorological observations.

During the summer in the year 1888, Messrs. Masato and Shimono were sent up Mount Gozaishodake for the meteorological study, and they made simultaneous observations on its summit and at Yokkaichi for thirty days. This expedition was the first high level observations in our country, extended so longer as one month at the altitude over one thousand metres.

In the year 1889, Messrs. Nakamura and Kondo were ordered to ascend Mount Fuji for the same purpose, and stayed five weeks on its summit and at Yamanaka, a temporary station established in the midway between the summit and the lower station at Numazu.

In the same year, Mr. Tonno, of Yamaguchi Higher Middle School, and his assistants made simultaneous observations on the mountain called Higashi Hôben and at Yamaguchi during three whole months, August, September and October; it was the first observations made in Japan so continued a long period at the height over seven hundred metres.

The last expedition was conducted on Mount Ontake in the year 1891, the members composed of Messrs. Masato, Kondo of the Observatory and some assistants joined by the Observatory and neighbouring stations; all the members passed six weeks on its summit or at the temporary station at Kurosawa.

The following table shows the principal ascents to the high mountains during which the regular meteorological observations were made.

Name of Mountains:	Altitude:	Year and Month:	Days of Observations:	Observers.
	Metres			
Fuji	3,718	3-5 August 1880	3	Messrs. Nakamura, Wada.
Fuji	3,718	4-6 September 1887	3	Knipping, Masato.
Gozaishodake	1,200	4 Sept.— 3 Oct. 1888	30	Masato, Shimono.
Fuji	3,718	1 Aug.— 7 Sept. 1889	37	Nakamura, Kondo.
Higashi Hoben	736	1 Aug.—31 Oct. 1889	92	Tonno and others.
Ontake	3,062	1 Aug.—12 Sept. 1891	43	Masato, Kondo and others.

II. TOPOGRAPHY OF STATIONS.

MOUNT FUJI.

This highest peak of Japan is situated near the SE coast of Honshu (greatest island of Japan named Nippon by foreigners) in the latitude of 35° 22′ N and the longitude 138° 44′ E. It is an extinct volcano (last eruption in 1707) and has therefore a regular conic form, isolated from other mountain ranges; its prominent point is elevated about 3,780 metres above the sea level. All these conditions rendered the execution of our purpose of high level investigations most convenient and suitable; in further it has much advantages as to its accessibility and comfort of the observers. There are several roads available for the ascent, requiring nearly seven or eight hours on foot, and on the top as well as on its slope, some stone huts or caverns are established for the help of tourists during the bad weather. On the southeastern side of the ancient crater we founded one dozen of stone huts where the tourists could make observations more or less satisfactory. During the observations of the year 1880, the lower station has been chosen at **Hara** in the southern direction, about 30 kilometres in a horizontal distance, but in the year 1882, the **Numazu** station, (in the direction SE, near the shore), was established and has been always selected for the comparing lower station in the subsequent expeditions. Besides this station, the members of the expedition in the year 1889 established a temporary middle station at Yamanaka in the eastern side, a horizontal distance of about 12 kilometres from the summit; Yamanaka is a small village situated on the high plain surrounding Fuji Mountain, at a height of nearly 1,000 metres above the sea level.

MOUNT GOZAISHODAKE.

This peak is situated near the western coast of Owari Bay in Central Nippon, 20 kilometres of distance from the sea, its summit 1,200 metres high is entirely bald, except some small plants or grasses. In this region, Gozaishodake is the most prominent peak from which all surrounding plains are splendidly visible in a fine weather; but no building having been found, the members of the expedition have passed more than four weeks in the canvas tent.

The lower station has been placed in the year 1888 at Yokkaichi, a horizontal distance of 20 kilometres in the southern direction from this peak, just on the coast of Owari Bay.

MOUNT ONTAKE.

The mount is situated in the middle part of Central Nippon, about 100 kilometres from the Sea of Japan and the Pacific Ocean; its highest peak is elevated about 3,100 metres above the sea level. Ontake is also extinct volcano, and some *solfatares* are found yet at places in its valleys. But the position selected for the meteorological observations having been on the top, no influence was effected by these subterranean disturbances.

There are many lower stations surrounding this mount in a circle of about 50 kilometres of radius; Gifu station is situated in a direction of SW at about 82 kilometres, the stations of Nagano, Hamamatsu, Kanazawa in the direction of E, S and N respectively at about 90—100 kilometres; in addition a temporary station was placed at Kurosawa in the adjacent valley at a distance of 15 kilometres and 500 metres below the summit.

MOUNT HIGASHI-HŌBEN.

Although the altitude of Hōben exceeds not 800 metres, the observations made on this mount are of great importance from several points of view. Hōben is situated in the western extremity of Nippon, in the hilly regions straitened by the Sea of Japan and the Inland Sea, so that the climatic difference is very peculiar with regard to the other mountain observations. On its foot is the station of Yamaguchi, founded privately by Mr. Tonno, which was served as the comparing lower station; the simultaneous observations were carried out during three whole months.

III. DESCRIPTION OF INSTRUMENTS.

It is perhaps useless to mention here the details of all the meteorological instruments employed in each expedition as well as those appropriated at the comparative lower stations, but it seems to be sufficient to state that all the instruments are made by the foreign makers of good reputation, like Messrs. Casella, Negretti & Zambra, Fuess, Richard frères and they are carefully compared with the standards of the Central Observatory, thus their corrections are ascertained in perfect accuracy.

The barometers used in the upper stations are those of Fortin's type, like in the lower stations, but at some lower stations, mercurial station barometers are supplied instead of them; with these one tenth of a millimetre or two thousands of an inch can be observed by the aid of vernier.

The thermometers, except the minimum thermometer and the minimum terrestrial radiation thermometer, are all mercurial, graduated directly on the stems into 0.°1—

0.°5 Centigrade; in some cases Fuess thermometers of the same accuracy were used for the dry and wet bulb or for extreme air temperature. In a few days, Arago's conjugated thermometers made by Alvergniat, were employed for some experiments on the solar radiation; also single black bulb thermometer in vacuo was observed for the same purpose. The Stevenson's screen being not convenient to carry up to the high mountains, temperature in shade has been observed by exposing the thermometers in a small cottage made by some pieces of wooden board or canvas for the protection of sun's rays or rain drops and for the easy ventilation of the air. But on account of the soil being not covered generally by the grass, we have been always in failure to protect entirely from the ground radiation; so that, the temperature given must be somewhat higher in the daytime, lower during the night.

The pluviometer used generally at our meteorological stations is cylindrical, two decimetres in diameter; it is placed in most cases in the grass field about one half metre above the ground. But at higher stations, no grass being obtained, the pluviometer is placed sometimes on the roof, or on the rocks; hence the amount measured with this instrument must be slightly influenced by the strong winds or other disturbances.

The anemometer employed is of Robinson's type, managed with automatic recorder; it is generally fixed at some prominent points, but sometimes it is mounted close the roof. Its correction has been obtained by comparing the instrument with the Observatory's standard anemometer.

IV. AIR PRESSURE.

All the barometric observations made at different higher stations are given with those of the corresponding lower stations in the Tables I of the Appendix, readings being reduced only to the temperature of freezing water, but not to sea level. The following Table gives the mean values for the whole months:

Stations	Altitude	Month	2h am	4h am	6h am	7h am	8h am	9h am	10h am	11h am	Noon	1h pm	2h pm	3h pm	4h pm	5h pm	6h pm	8h pm	10h pm	M. N.	Mean
	m	(1889)	mm	mm	mm	mm	mm	mm	mm	mm	mm	mm	mm	mm	mm	mm	mm	mm	mm	mm	mm
Fuji	3718	VIII	490.4	490.5	490.5	490.6	490.9	491.0	491.0	491.0	490.9	490.7	490.7	490.5	490.6	490.6	490.4	490.5	490.5	490.6	490.7
Yamanaka ..	990	VIII	677.9	—	677.4	677.7	677.7	677.9	677.9	677.7	677.4	677.3	677.2	677.0	677.1	677.1	677.2	677.0	677.8	676.9	677.5
Numazu.....	10	VIII	757.4	—	757.9	—	758.0	—	757.9	—	757.5	—	757.0	—	757.0	—	757.3	—	758.0	—	757.5
Ontake......	3062	(1890) VIII	531.4	531.3	531.6	—	531.8	—	532.1	—	532.1	—	532.4	—	531.8	—	531.8	532.1	532.0	531.7	531.8
Kurosawa ...	832	VIII	690.[illegible]	690.7	691.0	—	690.9	—	690.6	—	689.5	—	689.5	—	689.4	—	689.6	690.5	690.8	690.8	690.4
Gifu.........	15	VIII	758.4	—	758.5	—	758.7	—	758.5	—	757.9	—	757.3	—	757.6	—	757.1	757.5	757.8	758.2	757.9
Gozaishodake	1200	(1888) IX	660.5	—	660.4	—	—	—	661.2	—	—	—	660.5	—	—	—	660.6	—	661.0	—	660.7
Yokkaichi...	4	IX	760.0	—	760.2	—	—	—	760.7	—	—	—	759.7	—	—	—	759.7	—	760.5	—	760.1
Hōben	736	(1889) VIII	696.2	—	696.4	—	—	—	696.9	—	—	—	696.2	—	—	—	695.8	—	696.3	—	696.3
"	"	IX	697.4	—	697.2	—	—	—	698.0	—	—	—	697.3	—	—	—	697.1	—	697.8	—	697.4
"	"	X	699.7	—	699.6	—	—	—	700.3	—	—	—	699.2	—	—	—	699.6	—	700.1	—	699.8
Yamaguchi..	35	VIII	754.0	—	754.4	—	—	—	754.4	—	—	—	753.3	—	—	—	753.2	—	754.2	—	753.9
" ..	"	IX	756.4	—	756.6	—	—	—	756.7	—	—	—	755.4	—	—	—	755.7	—	756.6	—	756.2
" ..	"	X	759.9	—	760.3	—	—	—	760.4	—	—	—	758.3	—	—	—	759.4	—	760.2	—	759.8

We have drawn the curves of diurnal variation at some certain stations relating to the above Table for the month of August; this Diagram shows that:

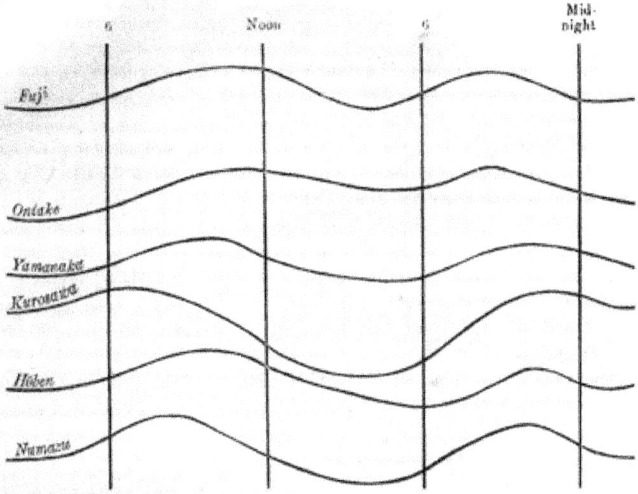

Curves of daily variation of air pressure during the month of August.

1. In the high level the variation of air pressure presents two maxima and two minima as in the low level;
2. The minimum of daytime is greater than those occurred during the night;
3. The absolute minimum occurs during the night at the height over 3,000 metres, while it appears in the daytime at lower station. From the Fuji observations, Mr. Nakamura concluded that this absolute minimum is more prominent during the clear days than the cloudy days;
4. The maximum and minimum of the night occurred nearly at the same time in higher and lower stations;
5. The time of occurrence of maximum and minimum in the daytime is later in the higher station than the lower.
6. The range of pressure is smaller in the higher station than the lower.

The difference between the absolute maximum and minimum in the mean pressure, during the month of August is shown in the following Table:—

Stations:	Height	Range of Pressure:
	m	mm
Fuji	3,718	0.6
Ontake	3,062	0.8
Yamanaka	990	0.8

Stations:	Height:	Range of Pressure:
Kurosawa	832 m	1.6 mm
Hōben	736	1.1
Numazu	10	1.1

Among them the diurnal ranges at Ontake and Kurosawa which are relatively more distant from the sea shore than others are influenced more specially by the air temperature.

On Mount Fuji the air pressure is reduced only to the 65th of that at sea level, 70th at Ontake, 87th at Gozaishodake, 89th at Yamanaka, 92th at Hoben. The rate of decrease in pressure diminishes rapidly with the altitude.

V. AIR TEMPERATURE.

The results of observations of air temperature in shade are collected in the Tables II of the Appendix, in which temperature is given in Centigrade degrees and the minus values are shown by adding 100 for avoiding the minus sign The following Table gives their mean values:

Stations	Altitude	Month	2h am	4h am	6h am	7h am	8h am	9h am	10h am	11h am	Noon	1h pm	2h pm	3h pm	4h pm	5h pm	6h pm	8h pm	10h pm	M.N	Mean
Fuji	3718	(1889) VIII	4.7	4.4	5.4	6.7	8.3	9.7	10.7	11.5	12.2	12.6	12.6	11.9	10.6	9.5	7.4	5.8	5.1	4.8	7.7
Yamanaka ...	990	VIII	18.0	18.1	19.9	21.4	22.4	23.5	24.5	24.8	24.8	24.2	23.5	22.6	21.7	20.0	20.1	19.6	19.1	18.1	21.5
Numazu	10	VIII	23.9	—	23.6	—	26.4	—	28.0	—	28.5	—	28.3	—	27.4	—	26.4	—	24.7	—	25.5
Ontake	3062	(1891) VIII	6.6	6.4	6.9	—	8.7	—	11.0	—	12.1	—	12.0	—	10.4	—	8.1	7.4	7.1	6.5	8.6
Kurosawa	832	VIII	16.4	15.8	15.9	—	19.3	—	23.4	—	25.7	—	25.9	—	24.5	—	22.0	18.9	17.8	17.0	20.3
Gifu	15	VIII	24.2	—	22.7	—	25.1	—	27.7	—	29.4	—	30.1	—	29.0	—	27.4	25.6	24.6	—	26.5
Gozaishodake	1200	(1888) IX	12.1	—	11.9	—	—	—	14.8	—	—	—	15.5	—	—	—	12.7	—	12.0	—	13.1
Yokkaichi ...	4	IX	18.4	—	17.6	—	—	—	22.1	—	—	—	24.5	—	—	—	21.6	—	19.1	—	20.7
Hoben	736	(1889) VIII	20.7	—	20.3	—	—	—	22.4	—	—	—	23.6	—	—	—	21.6	—	20.8	—	21.6
"	"	IX	15.6	—	15.4	—	—	—	17.5	—	—	—	18.3	—	—	—	16.3	—	15.6	—	16.4
"	"	X	11.8	—	11.0	—	—	—	13.3	—	—	—	14.4	—	—	—	11.7	—	11.2	—	12.2
Yamaguchi.	35	VIII	23.4	—	23.0	—	—	—	28.0	—	—	—	30.4	—	—	—	28.2	—	24.5	—	26.4
	"	IX	17.9	—	17.1	—	—	—	23.5	—	—	—	23.7	—	—	—	22.1	—	19.6	—	20.5
"	"	X	12.3	—	11.4	—	—	—	18.2	—	—	—	21.0	—	—	—	16.6	—	13.4	—	15.5

The absolute maximum and minimum observed during the ascents are:

Stations:	Max.	Min.
Fuji	21.4°	96.6°
Ontake	22.0	0.8
Gozaishodake	20.7	3.3
Yamanaka	31.1	11.8
Kurosawa	32.2	8.9
Hoben	29.1	0.4

The following Diagram indicates the diurnal variation of air temperature during the month of August.

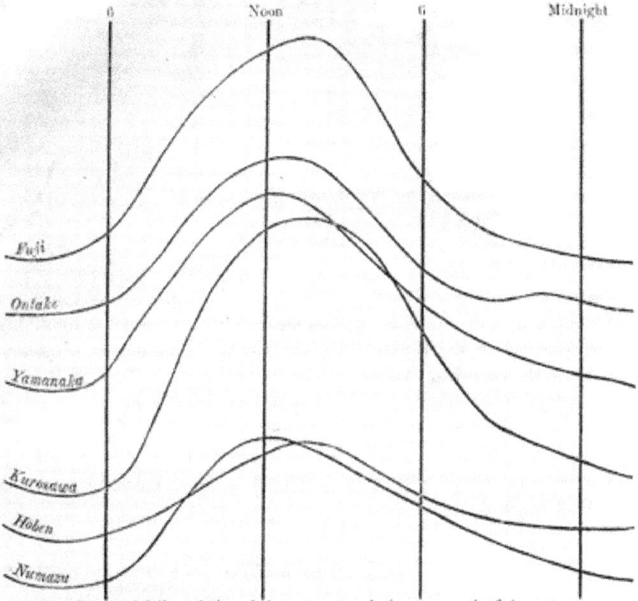

Curves of daily variation of air temperature during the month of August.

This Diagram shows us that:—

1. In the high level the minimum temperature occurs nearly at 4 am, the maximum at 2 pm, little before the respective times at lower stations;
2. The range between the maximum and minimum temperature is greater on the high stations than the lower; it increases also with the distance to the coast. The range during the month of August is as follows.

Stations:	Height:	Range:
	m	°
Fuji	3,718	8.2
Ontake	3,062	5.7
Yamanaka	990	6.9
Kurosawa	832	10.2
Hôben	736	2.7
Yamaguchi	35	7.4
Numazu	10	4.9

3. The rate of decrease in temperature with the height during the same month is as follows:

Higher Stations:	Lower Stations:	Difference of Height:	Rate of decrease per 100m:	Elevation for 1° of Decrease:
Fuji	Yamanaka	2,740m	0°.47	212m
"	Numazu	3,720	0.49	205
"	Hamamatsu	3,700	0.49	203
"	Tokio	3,710	0.49	204
Ontake	Kurosawa	2,220	0.50	191
"	Kanazawa	3,020	0.50	188
"	Fushiki	3,050	0.52	193
"	Hamamatsu	3,020	0.53	177
"	Gifu	3,030	0.58	172
Hōben	Yamaguchi	700	0.69	167

During the month of August the rate of decrease in temperature is smaller in the higher stratum than the lower, as shown in the above Table. From the observations made at Gozaishodake and Hōben during the months of September and October, we have obtained the following numbers:

Gozaishodake with Yokkaichi (September) 0°.62.
Hōben with Yamaguchi (September) 0°.63.
Hōben with Yamaguchi (October) 0°.50.

The rate of decrease in temperature at the same places is smaller during the colder months than the warmer.

VI. AIR HUMIDITY.

Absolute Humidity.—The tension of aqueous vapour, in millimetres, as computed from the observations of dry and wet bulb thermometers by Angot's tables, is given in the Tables III of the Appendix. The following Table shows the mean values:

Stations	Altitude	Month	2h am	4h am	6h am	7h am	8h am	9h am	10h am	11h am	Noon	1h pm	2h pm	3h pm	4h pm	5h pm	6h pm	8h pm	10h pm	M.N.	Mean
	m	(1889)	mm	mm	mm	mm	mm	mm	mm	mm	mm	mm	mm	mm	mm	mm	mm	mm	mm	mm	mm
Fuji	3718	VIII	5.0	4.5	5.8	5.3	5.6	5.8	5.7	5.7	5.6	6.0	6.2	6.2	6.3	5.9	5.5	5.5	5.2	5.2	5.5
Yamanaka ..	990	VIII	13.0	—	14.9	16.1	16.3	16.6	16.9	17.0	17.4	17.3	17.1	16.8	16.9	16.8	16.5	16.1	15.8	15.1	16.0
Numazu......	10	VIII	20.2	—	19.5	—	20.6	—	20.7	—	20.7	—	20.2	—	20.6	—	20.9	—	20.7	—	20.5
		(1890)																			
Ontake	3063	VIII	5.8	6.1	6.2	—	6.7	—	7.4	—	8.0	—	8.2	—	7.8	—	6.9	6.3	6.2	5.9	6.8
Kurosawa	832	VIII	13.9	13.2	13.1	—	14.4	—	15.2	—	15.1	—	15.3	—	15.4	—	15.8	15.2	14.7	14.1	14.6
Gifu........	15	VIII	20.1	—	18.7	—	18.9	—	19.6	—	19.4	—	19.8	—	19.7	—	20.1	19.7	19.7	—	19.5
		(1888)																			
Gozaishodake	1200	IX	9.7	—	9.5	—	—	—	10.7	—	—	—	11.2	—	—	—	10.6	—	9.8	—	10.2
Yokkaichi...	4	IX	14.6	—	14.1	—	—	—	15.5	—	—	—	15.2	—	—	—	15.6	—	14.5	—	14.8
		(1889)																			
Hōben	736	VIII	16.5	—	16.5	—	—	—	18.3	—	—	—	18.2	—	—	—	17.6	—	16.2	—	17.2
"	"	IX	12.7	—	11.5	—	—	—	12.6	—	—	—	13.6	—	—	—	12.7	—	12.2	—	12.4
"	"	X	7.8	—	7.5	—	—	—	8.2	—	—	—	8.1	—	—	—	8.3	—	7.8	—	8.0
Yamaguchi.	25	VIII	18.8	—	18.8	—	—	—	20.6	—	—	—	19.7	—	—	—	20.6	—	19.2	—	19.4
" ..	"	IX	14.0	—	13.0	—	—	—	14.6	—	—	—	14.7	—	—	—	14.7	—	14.3	—	14.3
" ..	"	X	9.8	—	9.5	—	—	—	10.8	—	—	—	9.6	—	—	—	10.4	—	10.2	—	10.0

From this table we obtain the following results:

1. The variation of the absolute humidity presents respectively two maxima and two minima just as the barometric curves, their times of occurrence differing little according to the height:
2. The absolute humidity decreases rapidly with the altitude, especially in the inland high stations. The rate of decrease in the month of August is as follows:

Higher Stations:	Lower Stations:	Difference of Height:	Decrease for 100m:
		m	mm
Fuji	Yamanaka	2,740	0.39
"	Numazu	3,720	0.40
"	Hamamatsu	3,700	0.43
"	Tokio	3,710	0.43
Ontake	Kurosawa	2,220	0.40
"	Kanazawa	3,020	0.37
"	Fushiki	3,050	0.39
"	Hamamatsu	3,020	0.45
"	Gifu	3,030	0.42
Höben	Yamaguchi	700	0.31

The absolute maximum and minimum observed in the above stations during the period of ascents are

Stations:	Max.	Min.
	mm	mm
Fuji	9.8	0.6
Ontake	10.2	0.1
Gozaishodake	15.7	5.2
Kurosawa	19.9	8.5
Yamanaka	21.1	9.8
Höben	22.9	3.5

Relative Humidity.—The relative humidity in percentage computed from the observations of dry and wet bulb thermometers with Angot's tables is shown in the Tables IV of the Appendix; the following Table gives their mean values:

Stations	Altitude	Month	2h am	4h am	6h am	7h am	8h am	9h am	10h am	11h am	Noon	1h pm	2h pm	3h pm	4h pm	5h pm	6h pm	8h pm	10h pm	M.N.	Mean
	m	(1889)	%	%	%	%	%	%	%	%	%	%	%	%	%	%	%	%	%	%	%
Fuji	3718	VIII	78	78	75	73	70	69	64	59	55	58	60	62	68	70	72	77	79	81	71
Yamanaka ..	990	VIII	96	—	96	92	86	82	79	75	74	74	76	79	83	86	90	94	95	96	89
Numazu.....	10	VIII	91	—	90	—	80	—	74	—	71	—	73	—	76	—	61	—	89	—	80
		(1891)																			
Ontake	3062	VIII	74	80	81	—	79	—	78	—	76	—	80	—	83	—	82	78	77	74	78
Kurosawa ..	812	VIII	96	96	96	—	84	—	70	—	62	—	62	—	67	—	90	92	95	96	83
Gifu.........	15	VIII	90	—	91	—	80	—	71	—	65	—	63	—	66	—	74	81	85	—	76
		(1888)																			
Gozaishodake	1200	IX	89	—	89	—	—	—	83	—	—	—	84	—	—	—	93	—	90	—	88
Yokkaichi .	4	IX	88	—	91	—	—	—	73	—	—	—	67	—	—	—	77	—	86	—	80
		(1889)																			
Höben .	730	VIII	93	—	93	—	—	—	89	—	—	—	84	—	—	—	91	—	89	—	90
"	"	IX	93	—	90	—	—	—	84	—	—	—	82	—	—	—	91	—	91	—	88
"	"	X	75	—	78	—	—	—	71	—	—	—	67	—	—	—	79	—	76	—	74
Yamaguchi.	35	VIII	87	—	89	—	—	—	68	—	—	—	61	—	—	—	79	—	84	—	77
"	"	IX	91	—	92	—	—	—	69	—	—	—	61	—	—	—	74	—	87	—	79
"	"	X	90	—	91	—	—	—	66	—	—	—	52	—	—	—	72	—	87	—	76

This Table shows that :

1. The diurnal variation of the relative humidity presents two maxima and two minima, more specially in the inland high stations.
2. The relative humidity increases with altitude within nearly one thousand metres and decreases with it outside this limit ;

From the daily observations it is assumed that the variation of the relative humidity on the high mountains is so changeable with other meteorological conditions, that sometimes it reaches the minimum percentage of less than 5, which are succeeded suddenly by the saturation ; on mount Ontake, the absolute minimum of relative humidity observed attained 2% during the night.

VII. AIR CURRENT.

Wind Direction.—The number of observations of wind direction under 16 points is given in the Tables V of the Appendix. The following Table shows the percentage of frequency in the wind direction at some stations :

Station	Height	N	NNE	NE	ENE	E	ESE	SE	SSE	S	SSW	SW	WSW	W	WNW	NW	NNW	Calm
Fuji	3,718	2	1	2	1	1	0	1	4	4	4	27	12	22	1	10	2	6
Ontake ..	3,062	4	1	1	0	0	0	0	1	4	7	15	13	29	11	10	4	0
Kurosawa ..	832	3	1	3	1	5	2	9	2	9	2	3	1	1	1	2	1	54
Numazu	10	1	2	11	15	9	10	5	1	1	1	16	12	9	4	2	0	1

The Table indicates that :

1. During the month of August Westerly winds are more frequent in the high stations than the other, while Easterly winds prevailed at the lower stations.
2. The mean direction computed with the corresponding number of observations for each direction at times of observations is

	2^h	4^h	6^h	8^h	10^h	Noon	2^h	4^h	6^h	8^h	10^h	Midnight
Fuji	$S_{52}W$	$S_{62}W$	$S_{65}W$	$S_{53}W$	$S_{70}W$	$S_{45}W$	$S_{63}W$	$S_{77}W$	$S_{75}W$	$S_{75}W$	$S_{68}W$	$S_{64}W$
Outake	$S_{42}W$	$S_{75}W$	$N_{86}W$	$S_{87}W$	$S_{67}W$	$S_{79}W$	$S_{72}W$	$S_{81}W$	$N_{87}W$	$N_{85}W$	$N_{74}W$	$S_{86}W$

On the high mountains it seems that the winds veered till about 8^ham, then backed till midday and are comprised between SW and NW.

3. The mean direction seems approaching to the W when the altitude increases and it runs about parallel to the isobars, while at lower stations it cuts them nearly with the angle of 45°.

Wind Velocity.—Tables V of the Appendix show also the wind velocity in metres per second deduced from the bi-hourly observations. The following is the mean value at each station :

Stations	Month	2h am	4h am	6h am	8h am	10h am	Noon	2h pm	4h pm	6h pm	8h pm	10h pm	Mid-night
Fuji ..	(1889) VIII	mps 9,4	mps 8,8	mps 7,7	mps 6,6	mps 6,1	mps 6,3	mps 5,5	mps 6,3	mps 5,6	mps 5,6	mps 8,1	mps 7,7*
Ontake	(1891) VIII	14,6	15,1	14,3	11,8	9,7	7,7	7,3	7,9	9,7	11,7	12,8	14,2
Kurosawa ..	VIII	0,2	0,4	0,3	0,5	1,6	2,1	2,2	2,2	0,9	0,3	0,3	0,4
Gozaishodake	(1888) IX	8,6	—	8,6	—	5,8	—	6,4	—	8,2	—	8,3	—
Hōben	(1889) VIII	11,8	—	11,3	—	10,6	—	10,6	—	11,0	—	10,7	—
"	IX	10,6	—	10,8	—	7,6	—	9,0	—	11,0	—	11,1	—
"	X	9,8	—	11,1	—	7,7	—	7,9	—	10,2	—	9,9	—

* The number refers to the mean of eighteen days only.

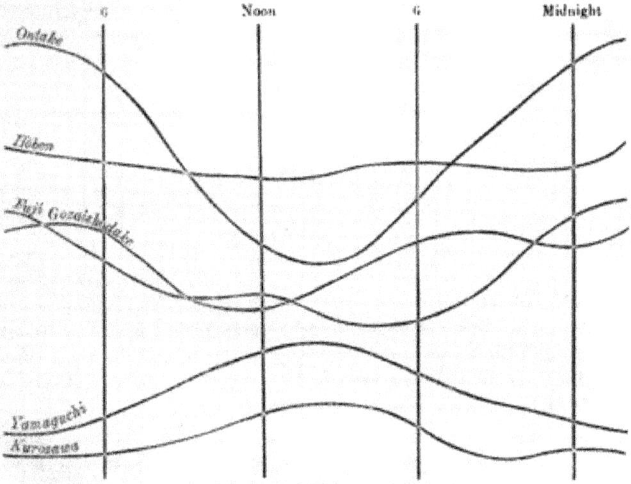

Curves of daily variation of wind velocity.

This Table shows that the wind velocity at high mountains reaches the maximum little before the sunrise, the minimum about 2hpm, so that the variation of hourly wind velocity is entirely reversed at the higher and lower stations.

For the excess of velocity at high mountains over that of corresponding lower stations, we have the following results:

Stations	Month	2h am	4h am	6h am	8h am	10h am	Noon	2h pm	4h pm	6h pm	8h pm	10h pm	Mid-night
		mps	mps	mps	mps	mps	mps	mps	mps	mps	mps	mps	mps
Ontake-Kurosawa	(1891) VIII	14.4	14.7	13.9	11.3	8.1	5.7	5.0	5.8	8.8	11.4	12.6	13.8
Gozaishodake-Yokkaichi	(1888) IX	5.5	—	6.7	—	2.8	—	2.2	—	4.6	—	4.9	—
Höben-Yamaguchi ...	(1889) VIII	10.1	—	9.6	—	7.1	—	6.2	—	8.0	—	8.7	—
" ...	IX	9.7	—	10.0	—	5.3	—	5.2	—	8.5	—	9.7	—
"	X	9.0	—	10.2	—	5.4	—	4.3	—	8.0	—	8.8	—

In the high stratum the velocity of wind exceeds more than ten times the velocity at lower station during the night, but three or four times only in the daytime.

VIII. PRECIPITATION.

The amount of rainfall measured at different high mountains and at corresponding lower stations is given in the Tables VI of the Appendix; the following Table shows its amount distributed to the hours of observations.

Stations	Month	2h am	4h am	6h am	8h am	10h am	Noon	2h pm	4h pm	6h pm	8h pm	10h pm	Mid-night	Total
		mm	mm	mm	mm	mm	mm	mm	mm	mm	mm	mm	mm	mm
Fuji	(1889) VIII	79	81	48	41	98	78	64	65	97	58	67	111	888
Yamanaka ..	VIII	27	—	72	23	32	58	54	47	62	50	94	61	680
Ontake	(1891) VIII	116	—	94	—	98	—	130	—	95	—	68	—	601
Kurosawa ..	VIII	22	39	22	43	12	11	25	15	18	6	20	33	266
Gozaishodake	(1888) IX	107	—	165	—	89	—	40	—	33	—	35	—	469
Yokkaichi ..	IX	19	—	34	—	43	—	13	—	7	—	6	—	122
Höben	(1889) VIII	9	—	2	—	2	—	4	—	8	—	0	—	25
"	IX	7	—	33	—	79	—	18	—	5	—	14	—	156
"	X	11	—	21	—	47	—	2	—	0	—	4	—	85
Yamaguchi..	VIII	6	—	6	—	2	—	3	—	6	—	3	—	26
"	IX	9	—	24	—	47	—	14	—	2	—	6	—	103
"	X	6	—	17	—	25	—	0	—	0	—	2	—	50

This Table shows that the amount of precipitation in the high stratum is greater than that of lower station, and except at Fuji and Yamanaka, the rainfall before noon exceeds that of afternoon.

The rate of increase in the precipitation varies between 11-15 mm. per 100 metres at the altitude over 3,000 metres, 27-29 mm. per 100 metres below 1,200 metres.

At the Fuji summit, during the storm of August 20th 1889, the rainfall in two hours rose to the enormous quantity of 53 millimetres; the greatest amount observed at Ontake attained 59 millimetres in four hours.

IX. AMOUNT OF CLOUDS.

The amount of clouds is observed at all the stations by estimation, the wholly clouded sky being noted as 10, the clear sky as 0. These results are exhibited in the Tables VII of the Appendix; the following Table shows the whole mean results:

Stations	Month	2^h am	4^h am	6^h am	8^h am	10^h am	Noon	2^h pm	4^h pm	6^h pm	8^h pm	10^h pm	Mean
		0–10	0–10	0–10	0–10	0–10	0–10	0–10	0–10	0–10	0–10	0–10	0–10
Fuji	(1889) VIII	4.3	4.0	4.4	4.2	4.4	4.4	5.3	6.4	6.4	5.1	4.6	4.8
Yamanaka ..	VIII	5.9	—	5.8	5.2	5.7	6.0	6.4	7.5	7.1	6.7	6.7	6.3
Ontake	(1891) VIII	6.0	6.0	6.4	6.7	7.2	8.1	8.5	8.2	6.5	4.4	5.4	6.7
Kurosawa ..	VIII	6.2	6.0	7.5	6.7	6.4	7.1	7.2	6.7	6.7	5.1	5.1	6.4
Gozaishodake	(1888) IX	7.0	—	6.4	—	6.6	—	7.1	—	7.4	—	5.7	6.6
Yokkaichi	IX	4.9	—	6.1	—	5.8	—	6.0	—	6.7	—	5.0	5.8
Hôben .	(1889) VIII	6.4	—	8.1	—	8.0	—	7.6	—	6.7	—	4.3	6.9
,,	IX	6.2	—	6.6	—	6.1	—	7.1	—	7.3	—	5.8	6.5
,, .	X	7.0	—	6.4	—	5.3	—	6.3	—	6.7	—	6.2	6.3
Yamaguchi	VIII	4.7	—	6.8	—	6.0	—	5.7	—	5.5	—	4.0	5.4
,, .	IX	4.6	—	6.1	—	5.6	—	7.2	—	6.1	—	4.6	5.7
,, ..	X	6.3	—	6.1	—	5.0	—	6.3	—	6.7	—	6.1	6.1

From this Table we conclude that the cloud amount increases with the height, but it seems decreasing when the altitude exceeds 3,500 metres above the sea level; and from the observations made at Hôben and Yamaguchi it appears that the cloudiness diminishes from August to October in the higher station, while in the lower station the case is just reversed.

We have drawn the curves of diurnal variation of the cloud amount at Fuji, Ontake and Hôben in the same month (VIII); these curves indicate that the time of maximum cloudiness occurs near 8^ham at Hôben, 2^hpm at Ontake, 5^hpm at Fuji and the time of minimum cloudiness appears near 8^hpm at Ontake, 11^hpm at Hôben, 4^ham at Fuji. In further, these curves show the excess of cloud amount in the daytime over that of the night, like in the lower stations.

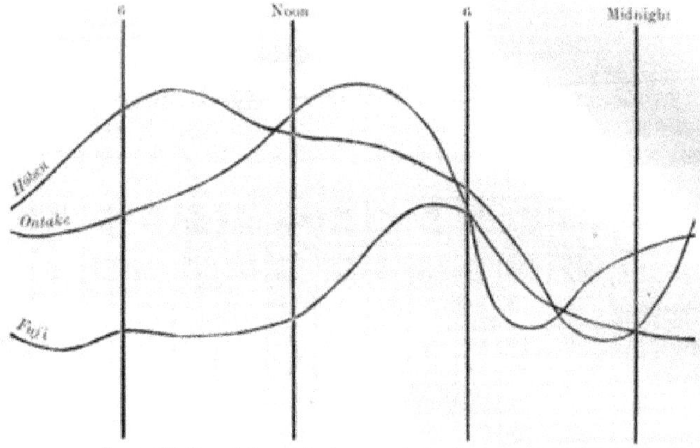

Curves of daily variation of cloud amount during the month of August.

X. ADDITIONAL OBSERVATIONS.

Solar Radiation.—During the ascent of 1887 on the summit of Fuji, some observations on the solar radiation have been made with Arago's conjugate thermometers. The difference between black and bright bulbs reaches the maximum of 11°.6 with the sun at nearly 40° from the zenith, the excess of bright bulb thermometer over the corresponding air temperature being 14°.8 and that of black bulb 26°.4.

At other high stations, similar experiments have been conducted with Kew radiation thermometers; the principal results are shown in the following Table:

Stations	Month	Absolute Maximum Solar Radiation Thermometer C°		Absolute Maximum Solar Radiation Thermometer in vacuo C°	
		Reading	Excess over the max. air temp.	Reading	Excess over the max. air temp.
Fuji	VIII	......	......	56.8	40.8
Yamanaka	VIII	......	......	72.8	43.6
Ontake	VIII	29.9	20.1	......	......
Kurosawa	VIII	47.5	15.3	......	......
Gozaishodake	IX	27.8	7.1	58.5	39.0

Terrestrial Radiation.—At each high station some observations on the terrestrial radiation have been made with Kew minimum terrestrial radiation thermometer; the results are shown in the following Table:

Stations	Month	Absolute Minimum Terrestrial Radiation Thermometer	
		Reading	Deficient from the minimum air temperature
Fuji	VIII	96.0°*	2.5°
Yamanaka	VIII	8.6	3.2
Ontake	VIII	95.2*	7.1
Kurosawa	VIII	5.3	3.1
Goraishodake	IX	98.5*	4.8

* Minus degrees added 100.

Earth Temperature.—During the expedition on Ontake, some earth thermometers were placed, one immediately below the surface of ground, other 30 centimetres below it; the respective results are given in the following Table:

Stations	Surface Temperature						Temperature below 30 cm.	
	2^h am	6^h am	10^h am	2^h pm	6^h pm	10^h pm	10^h am	10^h pm
Ontake	5.6°	5.5°	15.1°	17.8°	10.0°	6.4°	8.3°	9.1°
Kurosawa	16.6	16.5	33.2	36.3	23.8	18.0	....	

From these observations it seems that at high level the air temperature exceeds the ground surface temperature during the night and ranges below it during the day-time; the maximum ground surface temperature occurs near the noon, the minimum near 6^ham. But at lower station the ground temperature ranges always above the air temperature, during the night as well as the daytime. Below 30 centimetres of the ground the temperature ranges somewhat higher at 10^hpm than 10^ham, as observed at other lower stations.

XI. STORMS.

Storm of August 18th-21st 1889 on Fuji.—On the 18th of this month, a deep depression appeared on the S coast, traversed Shikoku on the morning of next day and proceeded northeastward into the Sea of Japan. During this storm, the barometer has indicated the following minimum pressure:

	Minimum	Time of Minimum	
	mm		
Fuji	482.9	7^h-8^hpm	20th August
Yamanaka	673.0	5 -6 pm	20th August
Numazu	754.6	4 pm	20th August

The rate of barometric fall has been only 2.7^{mm} in 24 hours at Fuji, which at Numazu it reaches 5-6 millimetres in the same interval. The pumping of mercury in the barometric tube was quick and so large on the summit that it has been very difficult to obtain the accurate reading.

The greatest wind velocity registered by the anemometer has been 36 metres per second; but the maximum velocity could not be obtained, the instrument having been destroyed by the wind. The rain fell extraordinary during this storm, the whole quantity amounting to 793 millimetres in 60 hours; the maximum rainfall reaching 69 millimetres in two hours from 10ʰpm on the 19th to the midnight.

Storm of August 26th 1889 on Fuji.—A shallow depression appeared on the morning of August 25th in Wᵗⁿ Kiushu and crossing slowly the Inland Sea went up to the Nᵗⁿ Sea of Japan. On the summit of Fuji, SW strong winds commenced blowing in the morning on the 25th, sending a dense fog which completely covered the top; at 2ʰam on the 26th the wind velocity attained 29 metres per second and increasing gradually it damaged again the anemometer. The change of air temperature was very small during both storms. The barometer continued falling on the 25th and attained the minimum on the 26th afternoon.

	Minimum (mm)	Fall in 24ʰ (mm)	Time of Minimum	
Fuji	483.5	4.7	4ʰpm	26ᵗʰ August
Yamanaka	670.9	3.2	3 pm	26ᵗʰ August
Numazu	751.5	3.2	2 pm	26ᵗʰ August

During this storm, the rate of falling mercury seems to be little greater on the top than in the lower station, but this rate is always influenced by the unconstancy of the mercurial top.

Storm of August 16-17th 1891 on Ontake.—On the 16th August, at 2ʰpm a deep depression appeared off Shikoku, then crossed the Wᵗⁿ Inland Sea, traversed the Northern Sea of Japan and passed Central Hokkaido. On the summit of Ontake the barometer commenced falling on the morning of the 16th, reached the lowest reading 523.1ᵐᵐ near 2ʰam on the next day; the minimum pressure observed at the lower stations are given with the corresponding readings at high stations:

	Minimum (mm)	Fall in 24ʰ (mm)	Time of Minimum	
Ontake	523.1	9.1	2ʰ30ᵐam	17ᵗʰ August
Kurosawa	683.5	8.2	4 am	17ᵗʰ August
Gifu	750.0	8.2	1 am	17ᵗʰ August

The fall of mercury on the summit is greater than Kurosawa and Gifu. The greatest wind velocity measured has been 45.2 metres per second, but the anemometer having been damaged by the wind, the absolute maximum velocity has not been obtained. The wind direction has been nearly tangent to the isobar. The rainfall during the storm has attained 171 millimetres in 24 hours from 6ʰpm on the 16th. The variation of air temperature during the storm has been very slight, as on the Fuji summit.

Storm of August 30th–31st 1888 on Gozaishodake.—A very deep depression came from Okinawa on the 29th, passed Kii Channel, traversed Central Japan and proceeded to the east extremity of Hokkaido on the 31st; Wakayama reported the lowest barometer 717.7^{mm} at 11^{h} 30^{m}pm on the 30th. On the summit of Gozaishodake the barometer commenced falling quickly from 9^{h}pm on the 30th and reached the lowest reading 641.9^{mm} at 3^{h} 30^{m}am on the next day.

	Minimum	Fall in 12^{h}	Time of Minimum	
	mm	mm		
Gozaishodake	641.9	12.7	3^{h}30^{m}am	31st August
Yokkaichi	740.1	11.7	3 am	31st August

The fall of barometer is little greater on the summit than the lower station. The greatest wind velocity measured by the anemometer before its damage has been 34 metres per second. The rain having been overflowed from the pluviometer, the total amount could not be obtained, but the quantity remained in the gauge attained 414.3^{mm} in 10 hours from the noon on the 30th. The variation of air temperature was very small, only 1 or 2° during the storm.

XII. ALTITUDE DETERMINATION.

Fuji.—Amongst several determinations of altitude of Fuji, the result obtained at the Surveying Bureau of the General Staff Office by means of triangulation is undoubtedly the most confident one; the chart published by the General Staff Office gives 3,778 metres above the sea level for its highest point. The 10th station where the barometer was placed being about 60 metres below it, the altitude of our station will be then

$$3{,}778 - 60 = 3{,}718 \text{ metres.}$$

The barometric observations made during the month of August 1889 applied to Sprung's Formula give for the same station.

Numazu	3,733
Hamamatsu	3,732
Tokio	3,733
Mean	3,733 metres

The result given by the barometric method during the month of August exceeds 15 metres that obtained by the triangulation, or about $\frac{1}{250}$ of the main altitude.

The variation of altitude with regard to the time of observations is computed for Numazu and their values are as follow:

	2^{h}	6^{h}	8^{h}	10^{h}	Noon	2^{h}	4^{h}	6^{h}	10^{h}
Altitude by Barometric Observations	3,690	3,699	3,733	3,756	3,766	3,767	3,749	3,723	3,703
Deviation from the True Altitude	−28	−19	+15	+38	+48	+49	+31	+5	−15

The deviation in excess is greatest at noon and 2^hpm near the time of maximum temperature, thence the altitude computed with the barometric observations at noon or 2^hpm is greatest and the deviation in deficit is least near 2^ham; the best result is that obtained by the observation at 6^hpm during the month of August.

Ontake.—Using the same formula, we have computed the altitude of Ontake from the simultaneous observations at some lower stations:

Gifu	3,056
Kanazawa	3,068
Nagano	3,046
Numazu	3,060
Mean	3,058 metres.

The altitude computed with Kurosawa's observation is 3,049 metres. Therefore the mean altitude is about 3,054 metres; the highest point being nearly 18 metres above the station, its altitude will be approximately 3,062 metres.

We have calculated the same altitude with the mean of bi-hourly observations; the result is:

	2^h	6^h	8^h	10^h	Noon	2^h	4^h	6^h	8^h	10^h	Mean
	3,055	3,054	3,066	3,083	3,096	3,102	3,099	3,079	3,061	3,051	3,075
Deviation from the mean	−20	−21	−9	+8	+21	+27	+24	+4	−14	−24	—

The greatest altitude corresponds to the time of maximum temperature and the result nearest the mean corresponds to that of 6^hpm, exactly the same as in the case of mount Fuji.

Gozaishodake.—Applying the observations made at Gozaishodake and Yokkaichi to Sprung's Formula, we have obtained as the altitude of Gozaishodake the following numbers:

	2^h	6^h	10^h	2^h	6^h	10^h	Mean
	1,195	1,195	1,209	1,208	1,198	1,196	1,200
Deviation from the mean	−5	−5	+9	+8	−2	−4	—

The most convenient time of observing the barometer for altitude determination during the month of September is nearly the same as the month of August.

Hōben.—The six daily observations continued during three months on Mount Hōben and at Yamaguchi applied to Sprung's Formula give the following numbers:

	2^h	6^h	10^h	2^h	6^h	10^h	Mean
August	732.3	733.6	737.0	734.1	735.3	732.5	734.1
September	730.7	731.1	735.5	735.5	727.7	729.4	732.2
October	730.4	731.3	734.9	735.4	730.0	728.7	731.8
Mean	731.1	733.0	735.8	735.0	731.0	730.2	732.7
Deviation from the mean	−1.6	+0.3	+3.1	+2.3	−1.7	−2.5	—

These numbers show that the altitude determined by the barometric observations decreases with monthly mean temperature and in the same month the greatest altitude falls near the time of maximum temperature; the mean value corresponds about the 6^ham or 4^hpm observations.

THE END

APPENDIX.

TABLE I. AIR PRESSURE.

Month & Year	Day	2h am	4h am	6h am	7h am	8h am	9h am	10h am	11h am	Noon	1h pm	2h pm	3h pm	4h pm	5h pm	6h pm	8h pm	10h pm	Mid-night	Mean
		FUJI 3718m.																		
		mm	mm	mm	mm	mm	mm	mm	mm	mm	mm	mm	mm	mm	mm	mm	mm	mm	mm	mm
VIII 1880	4	..	..	491.5	492.0	492.0	492.1	491.9	491.9	491.8	491.8	491.8	491.7	491.6	491.7	492.0	..	..	..	..
	5	..	..	492.2	491.8	492.2	492.2	491.9	491.8	491.7	491.9	491.7	491.7	491.5	491.5	491.5	..	..	..	..
IX 1887	4	..	..	..	..	..	..	..	..	..	488.8	488.7	488.6	488.7	488.7	488.7	488.9	489.1	489.1	..
	5	489.0	489.2	489.2	489.6	490.0	490.2	490.8	490.9	490.9	491.1	491.3	..	491.3	..	491.7	..	492.6	..	..
VIII 1889	1	492.3	492.6	492.8	..	493.4	..	493.6	..	493.5	..	493.5	..	493.5	..	493.6	494.0	494.2	494.1	493.42
	2	494.1	494.2	494.3	..	494.6	..	494.9	..	495.4	..	495.5	..	495.5	..	495.4	495.9	496.1	495.8	495.14
	3	495.7	495.7	495.8	..	496.0	..	496.1	..	495.5	..	495.3	..	494.7	..	494.3	494.2	493.4	492.8	494.96
	4	492.0	491.8	491.8	..	491.8	..	491.7	..	491.5	..	491.2	..	491.3	..	491.2	491.4	491.2	490.8	491.47
	5	491.0	491.0	490.9	..	491.1	..	491.3	..	491.1	..	491.0	..	490.8	..	490.9	491.0	491.1	491.1	491.02
	6	490.9	491.0	491.2	..	491.8	..	492.0	..	492.0	..	491.9	..	492.1	..	492.1	492.4	492.5	492.4	491.86
	7	491.8	491.7	491.7	..	492.3	..	492.4	..	492.4	..	492.3	..	492.2	..	492.5	492.6	492.8	492.2	492.24
	8	492.1	491.9	492.1	..	492.3	..	492.8	492.8	492.5	492.4	492.2	492.4	492.1	492.0	491.9	492.0	491.6	491.7	492.10
	9	491.4	491.4	491.9	492.0	492.5	492.4	492.3	492.5	492.2	492.1	492.0	491.8	491.7	491.6	491.6	492.1	492.2	492.0	491.91
	10	492.0	491.8	491.9	492.2	492.4	492.8	492.4	492.3	492.2	492.0	491.9	491.9	491.9	492.0	491.9	492.1	492.3	492.2	492.08
	11	491.7	490.9	490.8	491.0	491.1	491.4	491.1	491.8	491.9	491.9	491.8	491.7	491.8	491.8	491.8	492.1	492.2	492.2	491.62
	12	491.9	491.7	491.8	491.8	492.0	492.1	492.1	492.1	492.0	492.0	491.8	491.7	491.8	491.7	491.6	491.6	491.9	491.6	491.82
	13	491.2	490.8	491.1	491.3	491.2	491.2	491.3	491.3	491.1	490.9	490.9	490.7	490.8	490.8	490.8	491.1	490.9	490.8	491.00
	14	490.6	490.4	490.6	490.7	490.8	491.0	490.9	490.9	490.8	491.0	490.8	490.7	490.6	490.8	490.9	491.0	491.2	491.3	490.82
	15	491.1	491.1	491.2	491.2	491.4	491.4	491.7	491.8	491.7	491.7	491.8	491.9	491.9	492.0	492.0	492.3	492.6	492.5	491.77
	16	492.4	492.4	492.9	493.0	493.5	493.5	492.6	493.5	493.4	493.4	493.3	493.3	493.4	493.1	493.3	493.4	493.5	493.5	493.22
	17	493.1	493.1	493.2	493.2	493.5	493.4	493.6	493.3	493.4	493.3	493.2	492.8	492.8	492.8	493.0	492.8	492.5	492.0	493.02
	18	491.1	491.2	491.4	491.5	491.2	491.1	491.1	490.6	490.7	490.5	490.2	490.5	490.3	489.9	490.0	489.9	489.6	490.1	490.57
	19	489.1	489.7	488.2	488.1	488.7	488.3	488.7	488.5	487.6	487.6	487.6	485.6	485.8	485.3	486.5	485.6	485.8	484.5	487.30
	20	484.4	485.3	485.6	485.7	487.1	486.6	484.0	485.1	486.2	485.0	484.8	483.0	485.2	486.2	484.2	482.9	484.2	485.0	484.91
	21	484.6	486.3	486.1	486.4	488.5	489.0	489.5	489.6	489.8	489.7	489.6	489.6	489.9	490.2	490.3	491.0	491.2	491.1	488.99
	22	490.8	491.4	490.9	491.4	491.6	492.0	492.0	492.3	492.2	491.5	491.4	491.4	491.4	491.4	491.4	491.8	492.3	492.1	491.61
	23	491.6	492.3	492.4	492.4	492.4	492.6	492.8	492.5	492.6	492.4	492.3	491.9	492.3	492.3	492.2	492.5	492.5	492.4	492.36
	24	492.1	492.2	491.8	492.0	492.0	492.0	492.0	492.0	491.9	491.7	491.4	491.2	491.1	491.1	490.8	490.9	490.9	490.4	491.46
	25	489.6	489.3	489.6	489.6	489.6	489.8	489.7	489.5	489.3	489.2	488.7	488.6	488.2	488.2	488.0	488.1	487.5	487.2	488.73
	26	485.9	485.6	485.5	485.5	485.5	485.5	485.5	485.2	485.3	485.2	484.3	484.6	483.5	484.9	486.2	486.2	485.2	485.0	485.31
	27	486.0	486.2	486.6	486.8	487.1	487.0	487.0	487.1	487.2	487.1	487.0	487.1	487.1	487.1	487.4	487.5	487.6	487.3	487.00
	28	487.4	487.3	487.2	487.2	487.8	488.0	488.0	487.7	487.7	487.4	487.2	487.3	487.3	487.3	487.2	487.4	487.7	487.5	487.47
	29	487.3	487.2	487.4	487.6	487.8	487.9	488.1	488.1	488.0	488.1	488.0	488.1	488.4	488.2	488.1	488.3	487.9	487.7	487.85
	30	487.7	487.5	487.4	487.4	487.5	487.7	487.8	487.7	487.8	487.8	487.8	487.8	488.2	488.2	488.4	489.2	489.7	489.5	488.21
	31	489.7	489.8	489.9	490.0	490.0	490.2	490.3	490.2	490.2	489.9	489.9	489.6	489.9	490.0	490.0	490.3	489.9	489.3	489.93
	Mean	490.41	490.48	490.52	490.64	490.91	490.95	490.98	490.97	490.94	490.73	490.66	490.52	490.56	490.60	490.63	490.76	490.78	490.57	490.68
IX 1889	1	489.2	..	489.3	..	..	..	489.7	..	..	..	488.6	..	..	..	488.5	..	488.2	..	488.92
	2	487.6	..	486.7	..	..	..	485.7	..	..	..	486.7	..	..	..	487.0	..	487.9	..	486.77
	3	487.8	..	488.2	..	..	..	489.4	..	..	..	489.3	..	..	..	489.7	..	489.5	..	488.98
	4	489.9	..	490.3	..	..	..	490.9	..	..	..	490.3	..	..	..	490.4	..	490.6	..	490.40
	5	489.2	..	489.5	..	..	..	489.0	..	..	..	488.4	..	..	..	486.4	..	487.0	..	488.25
	6	484.8	..	485.0	..	..	..	483.9	..	..	..	485.7	..	..	..	485.5	..	487.0	..	485.32
	7	486.5	..	486.6	..	..	..	486.8	..	..	..	486.2	..	..	..	485.3	..	484.9	..	486.06
General Mean		489.94	490.48	490.04	490.64	490.91	490.97	490.42	490.97	490.94	490.73	490.13	490.52	490.56	490.60	490.06	490.76	490.25	490.57	490.42
		ONTAKE 3062m.																		
VIII 1891	1	531.8	531.6	532.1	..	532.6	..	532.8	..	533.3	..	533.1	..	533.2	..	533.2	533.4	533.6	533.3	532.83
	2	533.1	532.8	532.9	..	532.9	..	533.3	..	532.9	..	533.0	..	531.8	..	533.1	533.7	534.2	534.2	533.16
	3	534.0	533.8	534.3	..	534.4	..	534.3	..	534.1	..	533.5	..	532.7	..	531.6	532.3	531.9	531.3	533.17
	4	531.0	531.4	532.1	..	532.2	..	532.7	..	532.6	..	532.8	..	532.8	..	532.2	532.5	532.8	532.8	532.33
	5	532.7	532.8	533.1	..	533.6	..	533.6	..	533.9	..	533.5	..	533.7	..	533.7	533.9	533.9	533.8	533.51
	6	533.6	533.8	534.0	..	534.2	..	534.2	..	534.3	..	534.0	..	534.2	..	534.0	534.2	534.2	534.0	534.06
	7	533.8	533.7	533.7	..	534.0	..	534.2	..	534.2	..	534.1	..	533.9	..	534.0	534.0	534.2	533.7	533.96
	8	533.5	533.6	533.5	..	533.9	..	534.3	..	534.6	..	534.5	..	534.5	..	534.4	534.6	534.9	534.6	534.24
	9	534.3	534.0	534.1	..	534.4	..	534.4	..	534.1	..	533.8	..	533.5	..	533.1	533.1	532.8	532.4	533.67
	10	531.5	531.1	531.3	..	531.0	..	530.6	..	530.9	..	530.5	..	530.1	..	529.6	530.1	530.0	529.1	530.48
	11	528.6	528.4	528.5	..	528.9	..	529.4	..	529.4	..	529.6	..	529.8	..	529.9	530.3	530.4	530.0	529.43
	12	530.0	530.0	530.2	..	530.9	..	531.4	..	531.9	..	532.0	..	531.9	..	531.7	532.3	532.4	532.3	531.42
	13	532.0	532.0	532.0	..	532.3	..	533.0	..	533.1	..	533.1	..	533.1	..	533.0	533.0	533.1	532.9	532.72
	14	532.3	532.0	532.0	..	532.5	..	532.7	..	532.7	..	532.6	..	532.6	..	532.8	533.1	533.0	533.1	532.62
	15	533.1	532.4	533.1	..	533.2	..	533.4	..	533.5	..	533.7	..	533.6	..	533.4	533.7	533.7	531.4	533.18

Month & Year	Day	2h am	4h am	6h am	7h am	8h am	9h am	10h am	11h am	Noon	1h pm	2h pm	3h pm	4h pm	5h pm	6h pm	7h pm	8h pm	9h pm	10h pm	11h pm	Mid-night	Mean
HARA 3m.																							
		mm	mm	mm	mm	mm	mm	mm	mm	mm	mm	mm	mm	mm	mm	mm	mm	mm	mm	mm	mm	mm	mm
VIII 1880	4	..	..	759.8	760.2	760.3	760.4	760.5	760.4	760.3	759.8	759.5	759.8	759.7	759.6	759.8	..	..	..	..	..	..	..
	5	..	..	760.8	761.0	761.1	760.8	760.9	760.6	760.4	760.2	760.2	760.1	759.9	759.7	759.8	..	..	..	..	..	..	..
TOKIO 21m.																							
IX 1887	4	..	..	..	..	..	..	..	..	..	756.9	756.9	756.9	757.2	757.4	757.8	758.6	759.6	760.1	759.9	760.8	760.8	..
	5	761.0	761.8	762.4	763.9	763.2	763.6	764.1	763.5	763.2	763.3	763.1	..	763.0	..	763.8	..	..	..	763.2	..	..	..
YAMANAKA 990m.																							
VIII 1880	1	678.2	..	678.8	679.2	679.2	679.9	680.0	679.8	679.6	679.4	679.4	679.0	679.1	679.5	679.5	679.6	680.1	680.2	680.4	..	..	679.38
	2	680.6	..	681.2	681.7	682.1	682.0	682.0	682.1	682.0	681.8	681.8	681.7	681.6	681.7	681.8	681.9	682.1	682.4	682.4	..	..	681.63
	3	681.8	..	682.4	682.5	682.2	682.1	682.5	682.3	681.9	681.3	681.0	680.6	680.3	680.1	680.2	680.3	680.4	680.5	680.1	..	..	681.35
	4	678.2	..	678.2	678.2	678.1	677.9	677.8	677.3	676.5	676.6	676.6	676.6	676.5	676.5	676.5	676.4	676.7	676.7	676.7	..	..	677.33
	5	676.6	..	676.5	676.6	676.6	677.0	677.1	677.0	676.9	676.7	676.7	676.6	676.7	676.7	677.2	677.2	677.1	677.1	677.2	..	..	676.88
	6	677.0	..	677.5	677.6	677.6	677.7	677.4	677.7	677.6	677.2	677.2	677.1	676.8	677.1	677.4	677.5	678.2	678.4	678.5	..	..	677.50
	7	678.3	..	678.8	679.0	679.2	679.0	678.9	678.7	678.5	678.4	678.2	678.4	678.3	678.1	678.5	678.8	679.0	679.2	679.4	..	..	678.68
	8	679.2	..	679.1	679.4	679.8	679.8	679.8	679.5	678.7	678.9	678.7	678.5	678.3	678.3	678.2	678.2	678.7	678.9	678.8	..	..	678.97
	9	678.2	..	678.8	679.0	678.9	679.0	678.9	678.7	678.5	678.2	677.8	677.7	677.8	677.7	677.7	677.7	678.2	678.2	678.3	678.3	678.2	678.28
	10	678.0	..	678.5	678.8	678.8	678.8	678.8	678.5	678.6	678.2	677.8	677.5	677.5	677.8	677.8	677.9	678.0	678.6	678.6	678.1	678.1	678.25
	11	677.7	..	678.1	678.2	678.3	678.6	678.5	678.4	678.0	677.9	677.9	677.7	677.8	677.7	677.8	678.0	678.5	678.8	678.8	678.4	678.4	678.13
	12	678.1	..	678.7	679.1	679.1	679.2	679.0	678.5	678.5	678.5	678.4	678.1	678.3	678.2	678.2	678.3	678.6	678.9	679.1	678.9	678.6	678.58
	13	678.5	..	678.7	678.9	678.7	678.7	678.6	678.4	678.0	677.8	677.8	677.5	677.9	677.9	677.8	678.1	678.2	678.2	678.3	678.2	678.2	678.28
	14	677.7	..	678.3	678.6	678.7	678.7	678.4	678.8	678.7	678.6	678.2	678.1	678.1	678.0	678.2	678.3	678.5	678.8	679.0	679.0	679.0	678.50
	15	679.1	..	679.8	680.0	680.0	680.1	680.1	680.0	680.2	680.3	680.2	680.2	680.4	680.1	680.2	680.5	680.8	681.2	681.2	681.3	681.2	680.10
	16	680.9	..	681.4	681.2	681.4	681.6	681.9	681.7	681.6	681.4	681.2	681.0	681.0	681.1	681.2	681.3	681.8	681.8	681.7	..	681.5	681.38
	17	681.2	..	681.3	681.2	681.5	681.5	681.6	681.1	680.7	680.5	680.3	680.3	680.5	680.3	680.7	680.6	680.8	680.7	680.7	680.0	..	680.97
	18	679.6	..	680.0	679.9	679.9	679.8	679.8	679.4	679.2	679.0	678.4	678.3	678.2	677.8	677.6	677.5	677.7	677.8	677.8	676.9	676.7	678.87
	19	676.7	..	676.1	676.1	675.2	675.8	675.7	675.5	675.7	675.3	675.5	675.1	675.1	674.9	675.1	675.1	675.1	675.3	675.2	..	675.3	675.72
	20	674.7	..	673.6	674.0	674.2	674.4	674.2	674.2	674.2	673.7	673.4	673.5	673.2	673.0	673.0	673.2	674.0	673.9	674.1	673.8	673.4	673.83
	21	673.4	..	674.1	674.1	674.3	674.5	674.6	674.9	675.0	675.9	674.8	675.0	675.1	675.2	695.2	675.9	676.5	676.8	676.9	..	676.8	674.83
	22	676.8	..	677.1	677.5	677.5	677.7	677.6	677.6	677.4	677.2	677.0	677.0	677.0	677.0	677.2	677.1	677.9	678.1	678.6	..	678.7	677.38
	23	678.5	..	678.5	678.6	678.7	678.9	678.8	678.6	678.3	677.9	677.8	677.2	678.2	678.1	678.5	678.7	679.0	678.9	678.5	679.0	679.0	678.43
	24	678.5	..	678.4	678.4	678.4	678.4	678.8	678.3	677.8	677.4	677.8	676.7	676.7	676.3	676.2	676.1	676.7	676.6	676.7	676.7	675.8	677.73
	25	675.8	..	675.8	675.8	675.9	675.9	676.0	675.8	675.1	674.8	674.5	674.1	674.0	673.8	673.9	673.7	673.8	673.9	673.9	673.7	673.5	674.98
	26	673.0	..	672.5	672.2	671.1	671.9	671.8	671.7	671.5	671.2	671.0	670.9	671.2	671.3	671.4	671.5	672.1	671.8	671.7	671.1	670.8	671.90
	27	670.7	..	670.9	670.9	671.1	671.1	671.1	671.3	671.3	671.1	671.0	671.0	671.1	671.2	671.2	671.3	671.8	671.9	671.8	671.8	671.8	671.12
	28	671.8	..	672.5	672.6	672.8	673.1	673.2	673.0	673.0	672.9	672.7	672.9	673.0	673.0	673.2	673.3	673.5	673.5	673.2	673.2	673.2	672.73
	29	673.2	..	674.2	674.5	674.6	674.8	675.0	675.4	675.4	675.5	675.3	675.4	675.5	675.7	675.5	675.4	675.7	675.7	675.8	675.3	675.6	674.83
	30	675.6	..	675.7	675.8	675.8	676.0	676.3	676.0	676.1	675.9	675.9	676.1	676.2	676.7	677.0	678.0	678.0	678.2	678.6	678.6	678.6	676.52
	31	678.5	..	678.9	679.3	679.4	679.7	679.5	679.3	678.9	678.8	678.7	678.6	678.6	679.0	678.9	679.2	679.3	679.4	679.4	679.3	679.3	678.98
	Mean	677.23	..	677.35	677.76	677.71	677.85	677.86	677.73	677.52	677.33	677.19	677.06	677.10	677.06	677.19	677.21	677.64	677.75	677.19	676.95	676.90	677.48
IX 1880	1	679.2	..	679.1	..	..	..	679.3	..	..	..	678.6	..	..	..	678.6	..	..	..	678.4	..	..	678.87
	2	677.1	..	676.3	..	..	..	675.8	..	..	..	673.7	..	..	..	673.7	..	..	..	674.2	..	..	675.13
	3	674.4	..	674.9	..	..	..	676.1	..	..	..	676.0	..	..	..	676.4	..	..	..	678.0	..	..	675.97
	4	678.3	..	678.6	..	..	..	679.2	..	..	..	678.6	..	..	..	678.7	..	..	..	678.8	..	..	678.70
	5	677.8	..	677.1	..	..	..	677.2	..	..	..	675.6	..	..	..	675.3	..	..	..	675.1	..	..	676.37
	6	672.5	..	671.8	..	..	..	670.6	..	..	..	668.7	..	..	..	670.8	..	..	..	672.0	..	..	671.05
	7	672.4	..	673.1	..	..	..	674.6	..	..	..	673.6	..	..	..	673.8	..	..	..	674.7	..	..	673.70
General Mean		677.86	..	677.23	677.34	677.71	677.55	677.54	677.73	677.52	677.33	676.75	677.65	677.10	677.06	676.55	677.31	677.64	677.25	677.41	676.92	676.90	677.32
KUROSAWA 832m.																							
VIII 1891	1	691.5	691.8	692.3	..	692.2	..	691.5	..	691.1	..	690.5	..	690.6	..	690.9	..	691.9	..	692.6	..	692.6	691.62
	2	692.3	692.3	692.2	..	691.7	..	691.7	..	691.4	..	691.2	..	692.6	..	692.4	..	693.3	..	693.6	..	693.7	692.37
	3	693.6	693.5	693.6	..	693.6	..	693.3	..	692.8	..	692.6	..	691.7	..	691.6	..	692.2	..	693.4	..	692.0	692.77
	4	691.1	690.9	691.3	..	691.0	..	691.2	..	691.0	..	690.9	..	690.8	..	690.9	..	691.2	..	691.8	..	691.6	691.13
	5	691.1	691.6	691.7	..	692.4	..	691.9	..	691.7	..	691.0	..	691.5	..	691.8	..	692.4	..	692.9	..	692.7	691.93
	6	692.6	692.7	693.0	..	693.1	..	692.5	..	692.2	..	691.7	..	691.9	..	691.8	..	692.3	..	692.6	..	692.6	692.42
	7	692.6	692.5	692.6	..	692.4	..	692.0	..	691.8	..	691.0	..	690.7	..	690.9	..	691.6	..	692.0	..	692.2	691.82
	8	692.2	692.2	692.4	..	692.5	..	692.2	..	691.4	..	690.8	..	690.3	..	690.9	..	692.0	..	692.4	..	692.5	691.82
	9	692.5	692.3	692.6	..	692.5	..	691.7	..	690.5	..	689.6	..	689.3	..	689.3	..	690.0	..	690.2	..	690.1	690.88
	10	689.8	689.4	689.5	..	688.9	..	688.2	..	687.9	..	686.7	..	686.5	..	686.5	..	687.1	..	687.5	..	687.1	687.93
	11	686.9	686.5	686.9	..	687.1	..	686.9	..	686.5	..	686.0	..	685.7	..	686.0	..	687.0	..	687.5	..	687.6	686.75
	12	687.5	688.0	688.6	..	688.6	..	688.1	..	687.8	..	687.9	..	687.9	..	688.1	..	689.2	..	690.0	..	690.4	688.51
	13	690.2	690.2	690.6	..	690.3	..	689.7	..	688.9	..	688.7	..	688.6	..	688.9	..	689.5	..	690.1	..	690.2	689.67
	14	689.9	689.9	690.0	..	689.7	..	688.9	..	688.1	..	687.5	..	687.6	..	687.1	..	689.7	..	690.1	..	690.2	689.06
	15	690.1	690.5	690.8	..	690.7	..	690.3	..	690.0	..	689.8	..	690.0	..	689.9	..	691.0	..	691.8	..	691.8	690.54

TABLE I. AIR PRESSURE.

Month & Year	Day	2h am	4h am	6h am	7h am	8h am	9h am	10h am	11h am	Noon	1h pm	2h pm	3h pm	4h pm	5h pm	6h pm	8h pm	10h pm	Mid-night	Mean
ONTAKE 3062m.																				
		mm	mm	mm	mm	mm	mm	mm	mm	mm	mm	mm	mm	mm	mm	mm	mm	mm	mm	mm
VIII 1891	16	532.9	532.7	532.6	..	532.5	..	532.4	..	532.1	..	530.9	..	529.7	..	529.1	528.9	526.1	524.4	530.36
	17	523.8	524.6	525.8	..	525.6	..	526.7	..	527.8	..	528.7	..	528.9	..	529.2	529.7	530.3	530.0	527.65
	18	529.7	529.4	529.5	..	529.6	..	529.1	..	529.2	..	528.9	..	529.1	..	528.8	528.5	528.0	527.7	528.96
	19	528.1	527.9	528.5	..	529.0	..	529.6	..	530.3	..	530.4	..	530.2	..	529.8	530.0	530.3	530.1	529.52
	20	530.1	529.8	529.6	..	530.0	..	530.4	..	530.3	..	530.1	..	530.2	..	530.2	530.3	530.3	530.0	530.12
	21	529.4	529.1	528.9	..	528.9	..	528.7	..	528.6	..	528.7	..	528.2	..	527.8	528.5	528.3	528.5	528.59
	22	529.0	528.4	529.0	..	529.6	..	530.4	..	530.3	..	530.2	..	529.9	..	530.3	530.9	530.8	530.5	529.94
	23	530.2	530.1	530.7	..	531.0	..	531.3	..	531.3	..	531.8	..	531.6	..	531.7	532.5	532.8	533.0	531.50
	24	532.6	532.9	533.5	..	533.6	..	534.1	..	535.1	..	534.9	..	533.9	..	533.7	533.7	533.7	533.4	533.77
	25	533.2	533.0	532.8	..	532.8	..	533.5	..	533.2	..	533.1	..	533.0	..	533.1	533.3	533.1	532.8	533.07
	26	532.0	532.0	532.2	..	532.3	..	532.4	..	532.1	..	531.9	..	531.6	..	531.5	532.1	532.1	531.9	532.01
	27	531.5	531.4	531.7	..	532.2	..	532.4	..	532.5	..	532.5	..	532.3	..	532.4	532.8	532.9	532.7	532.27
	28	532.5	532.2	532.3	..	532.6	..	533.0	..	533.0	..	532.6	..	532.5	..	532.3	532.7	532.3	532.1	532.51
	29	531.7	531.6	531.8	..	531.6	..	532.1	..	532.1	..	531.6	..	531.4	..	531.6	531.7	531.4	531.1	531.64
	30	530.6	530.5	530.7	..	531.3	..	531.3	..	531.2	..	531.1	..	530.9	..	531.3	531.3	531.4	531.3	531.07
	31	531.0	531.2	531.6	..	532.1	..	532.6	..	532.3	..	532.2	..	532.4	..	532.6	533.4	533.3	533.2	532.32
	Mean	531.41	531.30	531.54	..	531.79	..	532.07	..	532.16	..	532.04	..	531.84	..	531.76	532.08	532.01	531.66	531.80
IX 1891	1	532.8	532.8	532.7	..	533.1	..	533.7	..	533.7	..	534.0	..	534.1	..	534.3	534.7	534.7	534.9	533.79
	2	534.3	534.4	534.5	..	534.9	..	535.3	..	535.3	..	535.1	..	535.2	..	535.4	535.7	535.5	535.7	535.11
	3	536.5	535.6	535.9	..	536.0	..	536.4	..	536.2	..	536.2	..	536.0	..	535.9	536.4	536.3	536.0	536.03
	4	535.7	535.6	535.6	..	535.6	..	535.8	..	535.5	..	535.6	..	534.7	..	534.6	534.7	534.2	533.8	535.07
	5	533.2	532.6	532.4	..	532.4	..	532.9	..	533.1	..	532.8	..	532.5	..	532.4	532.7	532.6	532.6	532.68
	6	532.5	532.3	532.3	..	532.6	..	532.8	..	532.8	..	532.7	..	532.5	..	532.6	532.6	532.6	532.4	532.56
	7	532.0	531.8	531.9	..	532.1	..	532.4	..	532.2	..	531.9	..	531.6	..	531.4	531.7	531.4	531.6	531.82
	8	531.5	531.1	531.2	..	531.4	..	531.6	..	531.4	..	531.1	..	530.9	..	530.6	530.4	529.9	529.5	530.88
	9	529.1	529.4	529.9	..	530.5	..	531.1	..	530.9	..	530.7	..	531.1	..	531.5	532.2	532.0	531.5	530.82
	10	531.2	530.4	530.4	..	529.7	..	530.4	..	530.0	..	530.9	..	530.9	..	531.4	531.8	531.8	531.7	530.88
	11	531.5	531.7	531.2	..	532.6	..	533.1	..	533.1	..	533.1	..	533.1	..	532.9	533.4	533.1	533.0	532.65
	12	532.7	532.9	533.4	..	533.9	..	533.6	..	534.1	..	533.6	..	533.3	..	533.3	533.5	533.3	532.7	533.36
General Mean		531.76	531.64	531.84	..	532.10	..	532.40	..	532.45	..	532.32	..	532.17	..	532.11	532.43	532.32	532.02	532.15
GOZAISHODAKE 1200m.																				
IX 1888	4	660.9	660.1	660.0	..	660.1	..	660.4	..	659.5	..	658.5	..	658.6	..	659.5	659.6	659.7	659.5	659.6
	5	659.1	658.8	658.8	..	659.2	..	659.7	..	659.4	..	659.0	..	658.8	..	658.5	659.3	659.7	659.8	659.2
	6	659.8	659.8	660.3	..	660.9	..	661.9	..	661.5	..	661.3	..	661.3	..	661.3	661.6	661.5	661.4	661.0
	7	660.8	660.6	660.8	..	661.0	..	661.1	..	660.6	..	660.1	..	660.4	..	660.4	660.6	660.9	660.8	660.6
	8	660.5	660.1	660.4	..	660.8	..	661.4	..	661.2	..	661.0	..	661.1	..	661.0	661.5	661.9	661.4	661.0
	9	661.3	660.9	661.5	..	661.4	..	661.1	..	660.8	..	660.3	..	660.2	..	659.6	661.2	661.3	661.1	661.0
	10	660.8	660.6	660.8	..	661.0	..	661.7	..	661.2	..	661.0	..	660.7	..	660.9	659.8	659.9	659.0	660.7
	11	658.8	658.8	658.9	..	659.2	..	659.1	..	658.4	..	657.4	..	657.5	..	657.0	657.0	656.5	655.0	657.8
	12	653.8	653.1	653.2	..	653.4	..	653.6	..	653.6	..	653.3	..	653.4	..	653.7	654.0	654.4	653.9	653.6
	13	653.7	653.8	653.9	..	654.4	..	654.5	..	654.2	..	653.2	..	653.6	..	654.7	656.2	656.6	656.3	654.7
	14	656.6	657.6	658.9	..	659.3	..	660.4	..	660.8	..	660.7	..	660.7	..	661.1	661.8	662.4	662.6	660.2
	15	662.6	662.5	662.8	..	663.2	..	663.6	..	663.5	..	663.3	..	663.3	..	663.6	663.9	663.8	663.5	663.3
	16	663.0	662.4	662.6	..	662.6	..	662.4	..	662.1	..	661.5	..	661.3	..	661.6	662.0	662.0	661.5	662.1
	17	661.0	660.9	661.0	..	661.2	..	661.8	..	661.5	..	660.7	..	661.1	..	661.4	661.7	661.5	661.5	661.3
	18	661.4	661.4	661.7	..	662.2	..	663.1	..	662.9	..	662.2	..	662.1	..	662.3	662.9	663.2	663.2	662.3
	19	663.1	662.9	663.0	..	663.2	..	663.7	..	663.7	..	663.2	..	662.4	..	662.3	662.9	662.2	661.8	662.9
	20	660.8	659.7	659.8	..	659.7	..	659.7	..	659.6	..	658.8	..	658.5	..	658.3	658.6	658.6	658.4	659.1
	21	658.6	658.6	658.7	..	659.3	..	659.7	..	659.6	..	659.4	..	659.7	..	660.1	660.3	660.5	660.1	659.6
	22	660.4	660.0	660.1	..	660.5	..	660.7	..	660.1	..	659.4	..	658.9	..	659.1	658.4	658.3	657.7	659.5
	23	657.5	657.7	658.8	..	659.5	..	660.2	..	660.6	..	660.9	..	661.0	..	661.7	661.8	662.5	662.4	660.4
	24	662.5	662.7	663.0	..	663.7	..	664.0	..	664.2	..	663.6	..	664.1	..	664.4	664.9	665.2	665.3	663.9
	25	665.4	665.5	665.6	..	666.4	..	666.8	..	666.7	..	666.3	..	666.2	..	666.2	666.6	666.6	666.3	666.3
	26	666.2	666.0	666.1	..	666.0	..	665.8	..	665.5	..	665.6	..	665.6	..	665.1	665.2	664.9	664.7	665.6
	27	664.3	663.9	663.9	..	663.6	..	663.7	..	663.4	..	662.5	..	662.3	..	661.9	661.3	662.0	661.7	662.9
	28	660.8	660.4	660.3	..	660.7	..	660.7	..	660.1	..	659.3	..	659.2	..	658.6	658.5	658.4	658.1	659.6
	29	658.0	657.9	658.1	..	658.4	..	658.5	..	658.0	..	657.7	..	657.8	..	658.1	658.5	658.7	658.8	658.2
	30	658.8	658.9	659.3	..	660.1	..	660.6	..	660.6	..	660.2	..	660.4	..	660.8	661.0	661.2	661.3	660.3
X 1888	1	661.0	660.8	660.8	..	660.6	..	660.5	..	660.0	..	659.7	..	659.8	..	659.8	659.7	660.1	660.2	660.2
	2	660.1	660.3	660.1	..	661.1	..	661.4	..	661.2	..	660.6	..	660.6	..	660.8	661.6	661.9	662.2	661.0
	3	662.3	662.5	663.0	..	664.0	..	664.4	..	663.9	..	663.7	..	663.9	..	664.0	664.6	664.8	..	663.7
	Mean	660.5	660.3	660.5	..	660.9	..	661.2	..	660.9	..	660.5	..	660.5	..	660.6	660.9	661.0	662.7	660.7

Month & Year	Day	2h am	4h am	6h am	7h am	8h am	9h am	10h am	11h am	Noon	1h pm	2h pm	3h pm	4h pm	5h pm	6h pm	7h pm	8h pm	9h pm	10h pm	11h pm	Mid-night	Mean
KUROSAWA 832m.																							
		mm	mm	mm	mm	mm	mm	mm	mm	mm	mm	mm	mm	mm	mm	mm	mm	mm	mm	mm	mm	mm	mm
VIII 1891	16	691.0	691.7	691.9	..	691.4	..	690.9	..	690.1	..	689.5	..	688.7	..	687.9	..	688.1	..	686.5	..	685.7	689.52
	17	684.2	683.5	684.1	..	684.8	..	685.7	..	686.1	..	686.3	..	686.4	..	686.5	..	686.7	..	687.1	..	687.6	685.75
	18	687.4	687.3	687.6	..	687.5	..	687.4	..	686.6	..	686.3	..	686.3	..	686.7	..	687.4	..	688.0	..	688.1	687.22
	19	688.2	688.3	689.1	..	689.1	..	688.9	..	688.5	..	688.3	..	687.9	..	688.0	..	688.8	..	683.5	..	689.6	688.68
	20	688.8	690.0	690.1	..	689.9	..	689.5	..	688.9	..	688.4	..	688.5	..	689.2	..	689.8	..	690.3	..	690.4	689.48
	21	690.0	689.8	689.4	..	689.7	..	689.2	..	688.0	..	687.5	..	687.4	..	687.4	..	687.4	..	687.4	..	687.4	688.38
	22	686.8	687.4	688.6	..	688.7	..	689.0	..	688.9	..	688.8	..	688.9	..	689.3	..	689.8	..	689.8	..	689.6	688.80
	23	689.7	690.2	690.0	..	690.2	..	690.4	..	689.3	..	690.7	..	690.7	..	690.9	..	691.9	..	692.4	..	693.0	690.78
	24	692.9	693.2	693.8	..	693.9	..	693.9	..	693.5	..	693.1	..	693.3	..	693.4	..	694.2	..	694.2	..	694.0	693.62
	25	693.9	693.8	693.9	..	693.8	..	693.4	..	692.4	..	692.0	..	691.9	..	692.2	..	693.1	..	693.0	..	693.4	693.07
	26	692.9	692.9	693.1	..	692.9	..	692.2	..	691.2	..	690.2	..	689.8	..	690.1	..	691.5	..	691.9	..	692.0	691.72
	27	692.1	692.1	692.5	..	692.4	..	692.0	..	691.2	..	690.6	..	690.6	..	691.2	..	692.4	..	693.0	..	693.1	691.93
	28	693.3	693.4	693.6	..	693.5	..	693.0	..	692.1	..	691.4	..	691.1	..	691.1	..	692.3	..	692.5	..	692.6	692.49
	29	692.4	692.2	692.4	..	691.9	..	691.2	..	690.2	..	688.8	..	688.5	..	688.9	..	689.8	..	690.0	..	690.0	690.52
	30	689.6	689.7	690.3	..	689.9	..	689.5	..	688.6	..	687.8	..	687.8	..	688.5	..	689.4	..	690.0	..	690.1	689.29
	31	690.6	691.0	691.6	..	691.4	..	690.8	..	689.7	..	689.4	..	689.1	..	689.8	..	691.2	..	691.6	..	691.9	690.68
	Mean	690.41	690.65	690.97	..	690.53	..	690.56	..	689.91	..	689.50	..	689.41	..	689.62	..	690.15	..	690.43	..	690.61	690.55
IX 1891	1	692.0	692.1	692.4	..	692.3	..	691.7	..	691.2	..	691.0	..	691.0	..	691.8	..	693.0	..	693.2	..	693.5	692.11
	2	693.1	693.4	693.4	..	693.5	..	693.2	..	692.2	..	691.9	..	692.2	..	692.5	..	693.6	..	693.8	..	694.1	693.07
	3	694.0	694.2	694.7	..	694.4	..	693.7	..	692.7	..	692.3	..	692.3	..	692.9	..	694.1	..	694.4	..	694.2	693.67
	4	694.1	694.0	694.2	..	694.1	..	693.8	..	692.9	..	692.0	..	691.6	..	691.1	..	692.8	..	692.6	..	692.7	692.99
	5	692.3	692.0	692.3	..	691.9	..	691.3	..	690.1	..	689.4	..	689.1	..	689.2	..	690.4	..	690.9	..	690.8	690.81
	6	691.0	691.2	691.2	..	691.2	..	690.5	..	689.5	..	689.2	..	689.2	..	689.7	..	690.5	..	691.0	..	691.1	690.44
	7	691.2	691.2	691.4	..	691.4	..	690.9	..	690.1	..	689.7	..	689.5	..	689.4	..	690.4	..	690.3	..	690.4	690.49
	8	690.2	689.9	690.1	..	690.0	..	689.9	..	689.6	..	689.1	..	688.9	..	689.0	..	689.8	..	689.6	..	689.0	689.60
	9	688.7	688.5	688.8	..	689.0	..	689.3	..	688.9	..	688.7	..	688.9	..	689.1	..	690.1	..	690.1	..	689.8	689.16
	10	689.7	689.5	689.7	..	689.4	..	689.1	..	688.9	..	688.9	..	688.6	..	688.9	..	689.6	..	689.7	..	690.1	689.34
	11	689.4	689.8	690.8	..	690.7	..	690.5	..	690.3	..	690.0	..	689.9	..	690.1	..	691.0	..	691.4	..	691.6	690.42
	12	691.4	691.8	692.0	..	692.3	..	692.0	..	691.0	..	690.0	..	689.6	..	690.2	..	691.0	..	691.3	..	690.9	691.12
General Mean		690.81	690.89	691.19	..	691.11	..	690.77	..	690.12	..	689.69	..	689.61	..	689.51	..	690.71	..	691.02	..	691.03	690.57
YOKKAICHI 4m.																							
IX 1888	4	758.2	..	757.7	..	..	..	757.8	..	..	..	756.0	..	..	..	756.4	..	..	..	757.5	..	..	757.3
	5	756.9	..	757.2	..	..	..	758.1	..	..	..	756.9	..	..	..	757.5	..	..	..	759.6	..	..	757.7
	6	759.7	..	760.5	..	..	..	761.0	..	..	..	759.8	..	..	..	759.9	..	..	..	760.7	..	..	760.3
	7	760.3	..	760.3	..	..	..	760.0	..	..	..	758.1	..	..	..	758.3	..	..	..	759.1	..	..	759.3
	8	758.8	..	758.9	..	..	..	760.1	..	..	..	759.1	..	..	..	758.9	..	..	..	760.3	..	..	759.3
	9	759.7	..	760.1	..	..	..	759.6	..	..	..	758.3	..	..	..	759.0	..	..	..	759.9	..	..	759.4
	10	759.5	..	759.5	..	..	..	760.1	..	..	..	759.1	..	..	..	758.8	..	..	..	758.4	..	..	759.2
	11	757.3	..	758.4	..	..	..	758.3	..	..	..	756.5	..	..	..	755.4	..	..	..	754.6	..	..	756.7
	12	751.3	..	750.5	..	..	..	750.7	..	..	..	750.1	..	..	..	751.3	..	..	..	752.4	..	..	751.0
	13	752.2	..	752.2	..	..	..	752.7	..	..	..	751.8	..	..	..	754.2	..	..	..	756.2	..	..	753.2
	14	757.1	..	759.2	..	..	..	760.7	..	..	..	759.8	..	..	..	760.5	..	..	..	762.2	..	..	759.9
	15	762.5	..	763.3	..	..	..	764.0	..	..	..	762.7	..	..	..	763.0	..	..	..	763.6	..	..	763.2
	16	763.1	..	762.9	..	..	..	762.4	..	..	..	760.9	..	..	..	760.8	..	..	..	760.8	..	..	761.8
	17	760.4	..	760.7	..	..	..	761.1	..	..	..	759.7	..	..	..	760.3	..	..	..	761.2	..	..	760.6
	18	761.0	..	761.9	..	..	..	762.5	..	..	..	760.7	..	..	..	761.1	..	..	..	762.9	..	..	761.7
	19	763.2	..	763.3	..	..	..	764.0	..	..	..	762.6	..	..	..	762.2	..	..	..	761.8	..	..	762.8
	20	760.4	..	759.4	..	..	..	758.9	..	..	..	756.8	..	..	..	756.3	..	..	..	756.7	..	..	758.1
	21	756.8	..	757.8	..	..	..	758.8	..	..	..	758.1	..	..	..	758.6	..	..	..	759.0	..	..	758.1
	22	759.2	..	759.1	..	..	..	759.4	..	..	..	757.9	..	..	..	757.1	..	..	..	755.9	..	..	758.1
	23	755.8	..	757.1	..	..	..	759.0	..	..	..	758.8	..	..	..	759.9	..	..	..	761.6	..	..	758.7
	24	762.1	..	763.6	..	..	..	763.5	..	..	..	763.2	..	..	..	764.4	..	..	..	765.6	..	..	763.6
	25	766.1	..	766.8	..	..	..	768.0	..	..	..	766.9	..	..	..	767.0	..	..	..	767.5	..	..	767.0
	26	767.2	..	766.8	..	..	..	766.2	..	..	..	764.8	..	..	..	764.6	..	..	..	764.9	..	..	765.7
	27	764.6	..	763.8	..	..	..	763.7	..	..	..	761.6	..	..	..	761.1	..	..	..	761.9	..	..	762.8
	28	760.1	..	760.6	..	..	..	760.0	..	..	..	756.6	..	..	..	757.6	..	..	..	758.3	..	..	758.8
	29	758.2	..	757.9	..	..	..	757.8	..	..	..	756.0	..	..	..	757.0	..	..	..	758.4	..	..	757.5
	30	758.8	..	759.7	..	..	..	760.8	..	..	..	759.8	..	..	..	760.5	..	..	..	761.5	..	..	760.2
X 1888	1	761.6	..	761.6	..	..	..	761.4	..	..	..	760.1	..	..	..	761.1	..	..	..	761.9	..	..	761.3
	2	762.0	..	763.5	..	..	..	763.9	..	..	..	762.3	..	..	..	763.5	..	..	..	765.1	..	..	763.4
	3	765.8	..	766.1	..	..	..	766.8	..	..	..	765.3	..	..	..	765.6	..	..	..	766.8	..	..	766.1
	Mean	760.0	..	760.3	..	..	..	760.7	..	..	..	759.3	..	..	..	759.7	..	..	..	760.5	..	..	760.1

TABLE I. AIR PRESSURE.

Month & Year	Day	2h am	6h am	10h am	2h pm	6h pm	10h pm	Mean
				HIGASHI HOBEN 736m.				
		mm	mm	mm	mm	mm	mm	mm
VIII 1889	1	698.9	699.2	699.8	699.6	699.7	700.6	699.63
	2	701.1	701.2	702.6	701.9	701.6	703.0	701.90
	3	703.1	703.6	704.4	703.1	702.0	702.2	703.07
	4	701.3	700.9	701.0	699.7	698.4	698.5	699.97
	5	698.4	698.0	698.5	697.5	697.0	697.8	697.87
	6	697.5	698.2	698.7	698.0	698.1	698.3	698.13
	7	698.1	698.9	699.5	698.6	698.3	699.0	698.73
	8	698.4	699.0	699.0	698.3	697.4	697.0	698.33
	9	698.0	697.8	698.4	697.6	697.5	697.9	697.87
	10	697.9	698.3	698.8	698.3	697.7	698.3	698.22
	11	698.6	698.5	699.1	698.8	698.1	699.3	698.73
	12	698.7	699.2	699.6	698.7	698.0	698.1	698.72
	13	698.2	697.9	698.2	697.6	696.7	696.7	697.55
	14	696.2	695.9	696.0	694.6	694.7	695.4	695.47
	15	695.2	695.2	695.1	695.7	695.7	696.3	695.55
	16	696.9	696.0	697.3	697.1	697.0	697.5	696.80
	17	697.6	698.1	698.5	697.7	697.1	697.2	697.70
	18	696.4	695.8	695.1	693.0	690.0	689.9	693.37
	19	688.3	686.9	686.6	685.4	686.0	687.7	686.82
	20	688.1	689.5	691.2	692.1	693.3	694.6	691.47
	21	694.7	695.9	696.8	696.7	696.6	697.6	696.38
	22	697.2	697.7	698.0	697.4	697.3	698.5	697.68
	23	698.3	698.1	698.7	698.0	697.7	698.4	698.20
	24	697.3	696.6	696.4	694.9	693.1	692.5	695.13
	25	691.3	690.5	690.4	690.0	689.7	690.1	690.33
	26	688.7	688.4	688.9	689.0	689.6	691.0	689.27
	27	690.6	691.3	692.1	691.8	691.5	692.5	691.63
	28	692.3	692.4	693.3	692.5	692.4	693.6	692.75
	29	692.8	693.5	694.5	694.0	694.3	695.7	694.13
	30	695.9	696.9	697.6	697.3	697.6	698.7	697.33
	31	698.2	698.2	698.5	697.4	696.9	697.4	697.77
	Mean	696.22	696.38	696.86	696.20	695.84	696.52	696.34
IX 1889	1	696.5	695.4	696.0	695.9	694.6	694.3	695.45
	2	693.3	692.4	693.2	693.6	694.4	695.2	693.68
	3	696.0	696.4	697.9	697.1	696.5	698.3	697.03
	4	697.9	697.9	698.5	697.4	697.2	697.0	697.65
	5	696.4	696.0	696.1	694.6	693.9	693.6	695.10
	6	692.2	692.7	693.5	693.0	692.8	693.9	693.02
	7	693.6	693.5	694.6	693.6	693.5	694.1	693.82
	8	694.3	694.4	696.0	695.3	695.1	695.9	695.17
	9	695.9	695.8	696.8	696.9	696.2	696.4	696.00
	10	696.2	695.8	696.8	695.7	694.8	694.4	695.62
	11	692.2	690.5	687.8	684.1	686.8	690.8	688.53
	12	692.9	694.5	696.3	697.1	697.2	697.9	695.98
	13	698.7	698.5	699.3	698.3	697.8	698.9	698.48
	14	697.9	697.8	698.6	697.5	697.4	698.0	697.87
	15	698.0	698.3	699.2	698.9	699.1	699.3	698.80
	16	699.5	699.5	700.8	699.9	700.0	701.1	700.13
	17	701.2	701.5	702.4	701.3	701.4	701.9	701.62
	18	702.0	702.3	703.0	701.7	701.7	702.9	702.27
	19	702.3	702.3	702.6	701.4	701.5	701.2	701.88
	20	700.4	699.4	699.9	698.6	698.6	698.9	699.30
	21	697.7	697.4	698.3	698.1	698.5	699.8	698.30
	22	700.0	700.4	701.1	700.6	700.2	701.0	700.55
	23	700.0	699.4	699.7	698.7	698.5	699.1	699.23
	24	698.7	698.6	700.0	699.0	699.5	700.3	699.35
	25	700.1	700.2	700.8	699.4	700.2	700.3	700.17
	26	700.0	699.8	700.5	699.2	699.2	698.8	699.58
	27	698.1	697.7	697.7	696.0	695.8	695.5	696.80

TABLE Ia. AIR PRESSURE.

Month & Year	Day	2h am	6h am	10h am	2h pm	6h pm	10h pm	Mean
				YAMAGUCHI 35m.				
		mm	mm	mm	mm	mm	mm	mm
VIII 1889	1	756.9	757.4	757.4	756.8	757.2	758.4	757.27
	2	759.1	759.5	760.1	758.6	758.9	760.8	759.50
	3	761.2	762.0	762.0	759.7	759.4	760.1	760.73
	4	759.4	759.0	758.1	756.4	755.5	756.2	757.43
	5	756.1	756.1	755.9	754.0	754.0	755.4	755.25
	6	755.4	756.1	756.0	754.7	755.0	756.0	755.53
	7	756.0	756.7	756.8	755.2	756.3	756.5	756.08
	8	756.3	756.9	756.3	755.0	754.5	755.6	755.77
	9	755.8	755.8	755.7	754.2	754.8	755.5	755.30
	10	755.7	756.4	756.1	754.7	755.0	755.9	755.63
	11	755.4	756.5	756.4	755.2	755.3	757.0	755.97
	12	756.1	756.9	756.9	755.2	755.2	755.6	755.98
	13	756.0	756.0	755.6	754.2	753.9	754.2	754.98
	14	754.0	753.6	753.8	751.4	752.2	752.9	752.98
	15	753.1	753.2	753.0	752.7	753.4	754.1	753.25
	16	752.7	753.9	755.2	754.1	754.7	756.4	754.50
	17	755.5	756.0	756.4	754.7	754.6	755.0	755.37
	18	754.2	753.9	753.2	749.8	747.5	747.5	751.02
	19	745.6	744.2	743.7	741.8	743.0	745.2	743.92
	20	745.6	747.2	748.2	748.7	750.5	752.2	748.73
	21	752.2	753.9	753.8	753.5	753.8	755.2	753.75
	22	754.8	755.5	755.1	754.3	754.6	756.1	755.07
	23	755.9	756.0	755.9	754.9	755.0	756.0	755.63
	24	755.0	754.6	754.2	754.9	750.7	750.2	753.25
	25	749.0	748.4	748.0	746.9	747.2	747.7	747.87
	26	746.4	746.3	746.5	745.9	747.1	748.6	746.80
	27	748.5	749.3	749.8	749.1	749.2	750.3	749.37
	28	750.2	750.4	751.0	749.8	750.0	751.3	750.45
	29	751.1	752.0	752.5	751.5	752.3	753.7	752.18
	30	754.3	755.4	756.6	754.8	755.6	757.2	755.48
	31	756.7	756.9	756.7	755.1	755.1	755.8	756.06
	Mean	754.04	754.32	754.38	753.14	753.24	754.24	753.90
IX 1889	1	754.9	754.4	753.9	753.7	752.5	752.5	753.65
	2	751.6	751.3	751.1	751.3	752.2	753.6	751.85
	3	754.4	755.1	755.5	754.5	754.4	756.5	755.07
	4	756.2	756.6	756.2	754.9	755.0	755.2	755.68
	5	754.5	754.4	753.8	752.1	751.6	751.7	753.02
	6	750.3	751.1	751.1	750.5	750.5	751.9	750.90
	7	751.9	752.0	752.3	751.0	751.3	752.1	751.77
	8	752.9	753.2	754.0	752.7	753.2	754.5	753.42
	9	754.6	754.7	755.2	753.4	753.3	755.0	754.37
	10	754.6	754.6	755.0	753.2	752.9	753.0	753.88
	11	750.5	747.9	745.6	741.4	744.7	749.3	746.57
	12	751.3	753.2	754.6	754.7	755.5	756.7	754.33
	13	758.1	758.0	758.2	756.5	756.5	757.4	757.47
	14	757.3	757.7	757.5	755.5	756.2	756.5	756.78
	15	757.6	758.4	758.0	757.2	757.8	757.8	757.80
	16	759.1	759.5	759.8	758.1	758.7	759.4	759.10
	17	760.7	761.5	761.3	759.4	760.2	760.4	760.58
	18	761.3	762.0	761.9	760.3	600.6	762.2	761.38
	19	761.5	761.9	761.5	760.1	760.3	760.4	760.95
	20	759.4	758.5	758.3	757.3	757.3	758.0	758.28
	21	756.6	756.8	757.4	756.8	757.2	759.0	757.30
	22	759.1	759.9	760.2	759.4	759.2	760.3	759.68
	23	759.6	759.4	759.2	757.9	757.5	758.5	758.68
	24	758.5	758.8	759.6	758.0	758.5	759.7	758.85
	25	759.9	760.4	760.4	758.6	759.2	759.6	759.68
	26	759.6	760.0	759.9	758.2	758.1	758.0	758.97
	27	757.5	757.5	757.4	755.4	754.7	754.7	756.20

TABLE I. AIR PRESSURE.

Month & Year	Day	2h am	6h am	10h am	2h pm	6h pm	10h pm	Mean
HIGASHI HOBEN 736m.								
		mm	mm	mm	mm	mm	mm	mm
IX 1889	28	695,1	695,2	695,7	695,0	695,4	695,8	695,37
	29	696,2	696,7	697,5	696,9	696,9	697,7	696,98
	30	698,0	698,3	700,6	700,2	700,5	702,0	699,93
	Mean	697,38	697,25	698,04	697,12	697,14	697,78	697,45
X 1889	1	701,8	702,4	703,3	701,9	701,8	701,9	702,18
	2	701,2	700,7	700,9	699,5	698,9	698,7	699,98
	3	698,0	697,8	698,6	697,7	698,4	699,3	698,30
	4	698,8	699,1	700,9	700,1	700,8	701,6	700,22
	5	701,1	701,9	702,3	700,3	700,1	699,9	700,93
	6	698,4	697,3	697,8	696,3	697,2	698,0	697,50
	7	698,0	698,1	699,4	697,9	698,5	699,8	698,62
	8	699,9	700,2	701,3	699,7	700,5	701,0	700,43
	9	700,6	700,7	701,7	700,7	700,8	701,9	701,07
	10	702,1	702,0	703,1	702,1	702,7	703,5	702,58
	11	703,5	703,7	704,7	703,0	702,8	703,8	703,58
	12	703,1	702,4	702,9	701,2	701,4	701,8	702,13
	13	701,7	701,7	703,3	702,1	702,4	703,2	702,40
	14	702,6	702,7	703,3	701,6	701,9	701,8	702,32
	15	700,0	698,7	697,8	695,9	696,5	695,1	697,17
	16	693,8	693,6	694,3	693,7	695,5	695,8	694,45
	17	696,6	697,1	698,1	696,9	698,0	698,5	697,53
	18	698,7	699,0	700,5	699,5	700,2	701,0	699,82
	19	701,0	701,7	702,8	701,3	701,5	702,5	701,80
	20	702,9	701,6	701,9	700,4	698,7	697,3	700,32
	21	695,0	693,0	693,1	693,5	694,1	695,7	694,12
	22	696,5	697,0	698,5	697,8	698,8	700,2	698,13
	23	700,4	700,5	701,3	700,2	700,3	701,8	700,67
	24	700,8	700,8	701,9	700,5	701,1	701,4	701,08
	25	700,9	701,0	701,9	700,4	701,2	700,8	701,03
	26	700,3	701,1	701,9	700,5	700,4	700,4	700,85
	27	700,8	700,6	701,1	699,2	698,7	698,1	699,75
	28	695,8	693,5	691,7	691,2	692,3	693,5	692,97
	29	694,1	696,3	698,0	698,3	699,4	700,2	697,72
	30	700,6	701,0	702,4	701,0	701,1	701,5	701,27
	31	701,3	701,6	702,2	701,7	702,1	702,9	701,97
	Mean	699,66	699,63	700,42	699,22	699,59	700,08	699,77

TABLE Ia. AIR PRESSURE.

Month & Year	Day	2h am	6h am	10h am	2h pm	6h pm	10h pm	Mean
YAMAGUCHI 35m.								
		mm	mm	mm	mm	mm	mm	mm
IX 1889	28	754,6	755,2	755,4	754,2	754,7	755,7	754,97
	29	755,7	756,7	757,2	756,1	756,3	757,6	756,60
	30	757,6	758,4	760,4	759,4	760,0	762,0	759,63
	Mean	756,38	756,65	756,74	755,30	755,67	756,64	756,25
X 1889	1	761,8	762,5	763,1	760,9	761,3	761,6	761,87
	2	761,2	760,9	760,5	758,2	758,4	758,6	759,63
	3	757,8	757,9	757,9	756,4	757,6	758,7	757,72
	4	758,7	759,3	760,1	758,7	759,8	761,0	759,60
	5	761,1	762,2	761,4	759,1	759,1	759,1	760,33
	6	758,0	757,2	757,0	755,1	756,3	757,2	756,80
	7	757,8	758,3	758,6	756,8	757,6	759,0	758,02
	8	759,9	760,7	761,0	759,1	760,1	760,9	760,28
	9	760,5	761,0	761,5	760,3	760,5	762,0	760,97
	10	762,1	762,7	763,0	761,5	762,3	763,6	762,53
	11	763,6	764,4	764,4	762,3	762,5	763,8	763,50
	12	762,9	762,8	762,7	760,5	760,9	761,7	761,92
	13	762,0	762,3	763,2	761,6	761,9	763,2	762,37
	14	763,0	763,4	763,4	761,3	761,3	761,5	762,32
	15	760,1	758,9	757,4	755,1	754,6	754,5	756,77
	16	753,8	753,7	754,1	753,2	754,9	755,7	754,23
	17	756,8	757,7	758,2	756,5	757,5	758,4	757,52
	18	758,5	759,3	760,1	759,0	760,4	761,3	759,77
	19	760,9	762,1	762,3	760,7	761,7	762,8	761,75
	20	761,8	761,9	761,6	759,7	758,9	757,3	760,20
	21	754,4	752,9	752,4	752,7	754,4	755,9	753,78
	22	756,7	757,8	758,4	757,5	759,4	760,9	758,45
	23	761,6	761,9	762,5	760,4	760,7	762,1	761,53
	24	761,6	761,7	763,0	760,6	761,4	762,0	761,72
	25	762,0	762,2	762,9	760,3	761,1	761,5	761,72
	26	761,4	762,3	762,9	760,4	760,5	760,9	761,40
	27	761,7	761,6	762,0	759,0	758,7	758,6	760,27
	28	756,5	754,0	751,5	751,1	752,6	754,4	753,35
	29	755,0	757,2	759,1	758,6	760,1	762,2	758,70
	30	762,5	762,8	763,8	761,4	762,1	762,4	762,50
	31	762,4	762,7	763,3	762,0	762,8	764,8	763,00
	Mean	759,94	760,20	760,43	758,71	759,41	760,24	759,82

TABLE II. AIR TEMPERATURE.

Month & Year	Day	2h am	4h am	6h am	7h am	8h am	9h am	10h am	11h am	Noon	1h pm	2h pm	3h pm	4h pm	5h pm	6h pm	8h pm	10h pm	Midnight	Mean
FUJI 3718m.																				
		°	°	°	°	°	°	°	°	°	°	°	°	°	°	°	°	°	°	°
VIII 1880	4	..	..	5.6	6.7	7.8	10.6	10.8	12.8	13.1	13.9	16.9	14.2	16.9	10.1	7.5	..	..	..	..
	5	..	..	2.9	4.0	4.3	4.6	4.9	5.1	5.4	7.2	7.2	7.9	7.2	6.9	6.0	..	..	..	..
IX 1887	4	..	..	..	..	..	..	..	..	..	..	..	..	2.7	0.6	0.0	99.1	98.8	99.4	..
	5	96.4	97.6	98.8	99.9	0.2	0.2	1.0	3.3	2.9	3.5	3.3	..	2.3	..	99.9	..	99.0	..	..
VIII 1889	1	4.5	4.2	5.4	..	10.0	..	12.3	..	15.5	..	16.2	..	13.6	..	10.9	7.6	6.2	5.6	9.23
	2	5.6	4.6	6.3	..	11.5	..	18.1	..	20.2	..	18.9	..	14.4	..	10.9	7.3	5.5	5.3	10.71
	3	5.4	4.8	5.9	..	10.6	..	12.4	..	14.1	..	15.8	..	12.4	..	8.5	5.6	5.4	4.8	8.81
	4	3.8	4.6	6.6	..	10.4	..	14.0	..	14.6	..	14.7	..	13.5	..	10.0	7.6	7.2	7.2	9.52
	5	5.2	5.4	8.2	..	14.4	..	18.7	..	17.3	..	17.9	..	14.8	..	10.2	7.2	6.0	4.9	10.85
	6	4.6	4.2	6.2	..	10.4	..	13.0	..	15.6	..	16.8	..	15.4	..	10.4	6.0	5.7	5.2	9.46
	7	6.0	6.0	6.9	..	7.7	..	9.2	..	10.8	..	13.4	..	11.5	..	7.2	6.3	5.8	6.2	8.00
	8	5.4	5.0	5.4	..	6.8	..	11.8	12.4	11.4	15.0	14.4	14.4	8.6	6.7	4.2	6.2	4.2	5.8	7.43
	9	5.6	4.6	5.8	5.6	7.6	8.9	10.0	11.8	14.4	16.1	15.9	12.6	8.6	6.8	6.4	6.0	6.0	5.4	8.02
	10	4.6	4.8	4.8	6.8	9.3	10.6	12.8	13.2	14.2	13.8	14.6	14.2	13.6	13.1	9.0	6.2	5.6	6.1	8.80
	11	5.6	5.2	4.4	5.7	7.2	9.0	10.0	12.0	12.6	14.4	14.5	12.8	12.2	9.2	8.2	5.8	5.7	4.5	7.99
	12	4.2	4.0	5.8	7.2	11.1	12.8	13.2	15.0	15.0	13.8	14.0	12.6	9.6	10.4	6.3	4.6	4.0	3.6	7.95
	13	3.6	3.6	3.8	4.4	5.6	9.2	9.8	11.6	13.0	13.2	12.9	7.4	7.9	7.6	5.9	4.4	3.1	3.3	6.41
	14	3.4	3.0	3.6	5.2	6.8	10.8	10.6	11.4	14.2	10.2	11.3	10.4	8.5	7.8	7.6	6.0	4.6	3.0	6.88
	15	1.6	0.8	4.5	8.2	10.6	12.2	14.0	15.1	15.8	15.4	16.2	16.8	15.3	14.2	9.3	6.0	4.8	4.7	8.63
	16	5.0	4.2	5.4	8.0	10.6	11.8	13.2	16.2	17.9	17.0	16.2	17.2	16.4	11.3	8.4	5.8	5.3	4.8	9.43
	17	4.4	3.6	6.0	10.2	12.6	14.2	15.2	16.0	15.7	16.0	15.4	16.1	14.0	10.9	8.6	5.5	4.9	2.8	9.06
	18	3.4	2.9	3.0	2.8	3.8	4.0	4.0	4.1	5.0	5.5	6.2	5.5	6.1	6.0	6.2	6.0	6.4	6.5	4.96
	19	6.8	6.0	5.4	5.8	7.1	6.8	6.4	6.1	5.8	5.4	5.1	4.9	4.8	4.5	4.3	4.7	5.1	5.3	5.57
	20	3.6	5.3	6.4	6.3	6.8	7.2	6.2	6.6	6.5	6.7	7.0	7.2	7.4	7.9	6.8	6.3	6.9	7.1	6.36
	21	7.6	7.5	7.1	7.5	7.1	7.4	7.1	7.2	8.0	10.6	9.0	9.6	9.6	8.1	7.3	5.8	5.8	4.6	7.21
	22	4.6	4.2	5.2	7.2	7.9	11.2	12.8	13.8	12.9	13.9	13.4	12.0	11.2	8.8	7.0	5.5	5.2	5.0	7.91
	23	5.6	4.8	5.4	5.8	8.2	10.7	13.0	14.0	14.8	14.4	13.0	14.0	11.6	7.2	4.9	5.4	5.2	5.0	8.07
	24	4.2	3.2	4.8	6.2	6.9	8.2	7.1	12.2	11.9	14.7	13.8	15.3	10.4	8.9	7.4	5.0	4.4	4.0	6.92
	25	4.2	4.0	4.4	5.4	6.8	8.9	9.6	10.6	11.4	11.6	10.6	10.4	7.9	6.7	5.5	4.4	4.7	3.9	6.45
	26	3.4	3.4	4.6	4.8	4.6	5.8	5.8	5.7	6.2	6.8	6.2	6.8	6.4	7.6	7.0	6.4	4.5	5.4	5.32
	27	4.6	5.2	6.6	6.8	8.4	9.7	10.3	11.0	10.2	10.6	10.6	10.8	10.3	10.7	9.4	7.4	5.8	7.2	8.00
	28	6.6	6.2	6.8	7.4	9.0	9.8	8.7	9.0	9.6	8.8	8.2	8.6	8.0	8.0	7.4	7.9	6.4	6.2	7.58
	29	6.6	6.4	6.4	6.6	6.8	7.8	8.8	9.6	9.3	9.8	9.4	8.2	8.6	7.2	6.8	5.6	5.0	5.0	7.06
	30	4.6	3.8	3.4	3.8	3.4	4.8	4.0	5.5	6.8	7.8	8.0	7.8	6.7	3.6	2.4	1.4	1.6	0.9	3.92
	31	0.8	2.2	2.4	4.6	6.8	3.6	9.1	3.4	8.0	8.6	9.8	9.0	8.1	5.7	4.3	3.6	2.4	0.6	4.84
	Mean	4.68	4.44	5.38	6.19	8.28	8.93	10.68	10.56	12.22	11.67	12.56	11.03	10.56	8.30	7.38	5.79	5.14	4.89	7.66
IX 1889	1	0.4	..	1.4	..	..	..	8.2	..	..	..	6.4	..	..	..	3.4	..	3.9	..	3.95
	2	4.6	..	5.0	..	..	..	6.7	..	..	..	5.8	..	..	..	5.2	..	3.9	..	5.20
	3	4.8	..	5.8	..	..	..	5.4	..	..	..	7.0	..	..	..	4.0	..	3.0	..	5.00
	4	3.0	..	3.9	..	..	..	7.4	..	..	..	7.4	..	..	..	4.0	..	4.0	..	4.95
	5	3.0	..	5.6	..	..	..	1.2	..	..	..	6.1	..	..	..	2.2	..	4.0	..	3.68
	6	6.2	..	6.6	..	..	..	5.8	..	..	..	6.4	..	..	..	1.6	..	3.0	..	4.93
	7	2.6	..	5.0	..	..	..	9.4	..	..	..	13.6	..	..	..	3.2	..	3.2	..	6.17
General Mean		4.46	4.44	5.26	6.19	8.28	8.93	9.87	10.56	12.22	11.67	11.63	11.03	10.56	8.30	6.61	5.79	4.95	4.80	6.25
ONTAKE 3062m.																				
VIII 1891	1	6.6	5.0	5.1	..	8.4	..	11.4	..	15.9	..	18.7	..	14.9	..	9.2	7.1	8.1	8.1	9.87
	2	8.2	8.5	9.2	..	9.3	..	10.0	..	9.1	..	9.2	..	7.1	..	7.3	6.4	8.0	8.4	8.39
	3	8.2	9.0	8.8	..	8.4	..	8.8	..	9.3	..	9.0	..	8.2	..	6.2	7.1	7.2	7.1	8.11
	4	8.1	8.0	8.9	..	9.2	..	9.9	..	9.1	..	9.0	..	9.2	..	7.1	6.0	6.5	8.4	8.28
	5	8.0	8.2	8.2	..	9.7	..	10.0	..	12.8	..	11.2	..	9.0	..	8.9	8.3	8.0	8.0	9.19
	6	8.6	8.4	8.0	..	8.0	..	7.0	..	10.8	..	11.1	..	11.3	..	9.1	8.8	7.7	7.9	8.89
	7	8.2	8.4	7.8	..	11.8	..	13.7	..	16.4	..	16.6	..	14.8	..	10.2	8.6	7.6	6.8	10.91
	8	6.4	6.2	8.1	..	11.7	..	13.4	..	14.0	..	15.4	..	14.0	..	11.0	9.2	8.3	8.1	10.48
	9	8.9	8.4	8.4	..	10.0	..	15.8	..	14.6	..	14.2	..	12.2	..	9.8	8.2	8.2	8.4	10.59
	10	8.0	8.3	8.2	..	8.8	..	8.8	..	10.2	..	10.8	..	9.8	..	8.6	8.5	7.9	8.1	8.83
	11	7.0	5.1	6.3	..	8.4	..	8.8	..	9.3	..	9.6	..	9.7	..	9.0	7.2	8.1	7.3	7.98
	12	6.2	5.4	6.3	..	7.8	..	13.8	..	15.4	..	15.4	..	12.4	..	9.0	8.2	8.0	9.0	9.74
	13	9.0	8.1	8.3	..	8.2	..	11.2	..	16.9	..	15.6	..	14.6	..	11.2	9.4	8.5	8.4	10.78
	14	8.4	8.1	8.5	..	11.5	..	15.7	..	18.0	..	18.2	..	15.1	..	11.7	10.2	9.4	9.4	12.02
	15	8.7	8.3	8.9	..	11.8	..	13.3	..	15.9	..	15.2	..	12.2	..	9.2	8.3	8.2	6.2	10.52

Month & Year	Day	2h am	4h am	6h am	7h am	8h am	9h am	10h am	11h am	Noon	1h pm	2h pm	3h pm	4h pm	5h pm	6h pm	7h pm	8h pm	9h pm	10h pm	11h pm	Mid-night	Mean
										HARA 8m.													
VIII 1880	4	° ..	° ..	° 22.2	° 24.9	° 26.1	° 26.8	° 27.6	° 28.8	° 30.0	° 30.3	° 30.0	° 31.1	° 29.6	° 29.2	° 27.9	° ..	° ..	° ..	° ..	° ..	° ..	° ..
	5	..	..	23.4	23.9	26.2	28.6	29.4	30.9	31.8	33.1	25.6	27.3	28.8	30.3	28.5	..	..	..	..	..	..	..
										TOKIO 21m.													
IX 1887	4	..	..	..	..	..	..	..	..	..	..	..	..	22.3	22.5	22.2	21.6	21.2	21.0	20.6	20.6	20.3	..
	5	20.1	17.6	17.8	18.6	20.3	22.2	22.5	23.3	25.7	25.8	25.3	..	25.6	..	23.3	..	..	..	20.2	..	..	..
										YAMANAKA 990m.													
VIII 1889	1	18.7	..	18.7	21.6	24.1	25.0	25.2	27.4	27.6	27.0	27.1	27.0	25.8	25.0	23.8	22.7	21.5	20.7	19.9	..	..	22.23
	2	18.6	..	18.6	21.3	23.1	26.0	25.9	28.9	28.7	29.7	28.7	26.9	26.7	24.9	23.9	22.2	20.6	20.1	19.9	..	..	22.77
	3	18.7	..	18.7	21.4	23.7	25.2	27.4	28.2	29.6	30.3	30.1	29.2	28.0	26.5	25.5	23.3	22.3	21.9	21.2	..	..	23.60
	4	19.3	..	19.4	21.9	23.6	25.7	26.9	28.1	28.1	25.5	24.7	24.8	25.2	23.9	22.6	21.5	21.1	20.5	19.1	..	..	22.00
	5	17.6	..	17.8	21.8	23.3	22.5	24.5	26.1	26.3	26.3	26.3	23.9	22.9	22.5	21.7	20.2	20.2	19.2	18.3	..	..	21.03
	6	17.0	..	17.6	20.2	22.8	24.5	26.5	28.1	28.6	28.0	26.9	26.6	25.9	24.5	22.7	21.8	21.6	21.6	20.7	..	..	21.90
	7	17.7	..	19.7	21.5	23.2	25.3	26.5	27.7	29.3	28.0	27.3	25.5	24.4	23.0	21.6	21.1	20.7	20.4	19.9	..	..	22.12
	8	18.8	..	18.9	20.8	22.2	23.7	25.1	26.0	27.5	25.2	23.7	25.3	23.5	21.1	20.9	20.0	20.0	19.9	19.6	..	..	21.17
	9	19.3	..	19.8	20.2	22.3	22.9	25.0	25.7	26.5	27.5	26.1	25.0	22.7	22.5	21.3	20.6	20.1	20.1	20.5	20.5	19.9	22.00
	10	19.0	..	18.6	21.4	23.5	25.1	24.7	26.1	26.5	27.7	28.0	26.5	24.7	22.8	22.3	22.1	21.4	21.7	20.9	20.6	21.2	22.25
	11	19.9	..	20.9	21.8	24.1	23.9	25.9	25.9	25.9	26.1	26.3	25.5	24.0	22.9	22.2	22.3	22.0	21.5	21.4	20.5	19.7	22.77
	12	19.9	..	20.5	22.4	22.7	23.8	25.9	27.1	29.1	27.6	26.9	26.9	23.9	24.0	22.1	21.1	20.3	24.1	20.5	19.9	19.7	22.63
	13	18.6	..	19.7	21.3	23.2	24.6	25.7	26.3	26.9	27.5	24.1	23.8	23.2	22.1	21.3	20.7	19.9	18.7	17.8	18.1	17.3	21.20
	14	15.4	..	16.8	20.2	22.6	22.6	23.3	25.4	25.1	25.6	24.7	23.2	23.4	21.0	20.5	20.0	19.5	19.4	18.9	18.3	17.0	19.93
	15	15.8	..	16.0	19.4	22.3	23.4	23.7	25.0	26.3	23.5	23.5	23.9	22.8	21.6	20.1	17.3	16.0	15.7	15.3	13.9	13.2	19.07
	16	12.2	..	15.0	17.9	21.3	23.0	24.0	24.5	26.1	24.6	24.0	23.9	23.0	21.7	20.3	19.5	18.6	18.6	18.0	..	17.2	18.92
	17	16.4	..	15.3	19.3	21.6	24.3	24.2	26.7	27.7	28.5	27.6	27.1	24.1	22.5	21.4	20.7	20.5	20.1	20.1	..	..	20.83
	18	19.3	..	18.8	19.2	19.4	19.0	20.3	22.6	19.7	18.9	19.1	19.3	18.9	20.3	20.5	21.0	20.1	20.1	20.7	20.7	20.7	19.78
	19	20.0	..	19.9	20.1	21.0	20.8	20.9	20.7	20.3	21.2	18.8	20.0	21.1	20.2	19.3	19.1	19.5	19.6	19.6	..	19.3	19.75
	20	19.2	..	19.3	19.6	19.3	19.7	20.1	20.3	19.5	20.1	20.2	19.6	20.1	20.3	19.5	20.0	19.9	20.0	20.0	19.9	19.8	19.73
	21	19.7	..	20.1	20.5	21.4	22.0	23.4	24.4	25.0	25.2	24.7	23.3	23.3	22.0	21.7	21.3	20.7	20.8	19.7	..	18.5	21.55
	22	18.7	..	18.7	21.3	23.1	24.1	24.9	26.5	27.3	27.4	27.1	24.3	23.7	23.4	22.6	21.2	20.9	20.7	19.3	..	19.3	21.88
	23	19.5	..	17.6	20.1	22.1	24.3	25.6	26.8	28.6	28.1	28.1	24.9	23.1	21.3	20.3	20.1	19.9	19.7	19.7	19.5	18.7	21.80
	24	17.1	..	16.4	19.2	21.5	23.7	24.5	26.0	26.5	27.5	25.7	25.5	23.2	24.0	22.3	21.3	20.1	19.5	19.0	19.9	..	20.83
	25	18.7	..	18.9	20.7	21.7	22.4	23.0	25.3	25.7	24.5	23.5	20.9	21.7	22.0	21.1	20.9	20.5	19.9	19.5	19.7	19.5	20.78
	26	18.5	..	18.5	20.0	20.3	21.7	23.1	22.3	24.1	23.9	24.1	21.3	20.3	19.3	19.0	18.7	18.7	18.7	19.5	19.5	19.7	20.45
	27	19.5	..	18.8	19.5	19.8	20.1	21.0	20.5	21.3	21.3	21.3	21.5	21.9	21.5	21.0	20.7	20.7	20.9	20.7	20.7	20.5	20.38
	28	20.8	..	19.4	19.5	19.9	20.1	20.1	19.5	19.9	19.5	19.3	19.4	18.5	18.4	17.9	17.2	17.1	16.9	17.0	16.6	16.3	18.98
	29	15.4	..	14.8	14.2	14.5	15.1	15.5	15.7	15.7	15.5	16.0	15.9	15.9	15.6	14.8	14.1	14.3	14.4	14.5	14.7	14.0	15.17
	30	13.9	..	14.1	14.3	14.5	15.3	15.3	17.1	17.7	18.1	17.9	18.0	17.9	16.9	16.0	15.2	15.1	15.1	15.2	14.5	13.9	15.40
	31	13.4	..	13.0	14.0	14.3	15.6	18.1	18.8	19.1	20.1	18.7	18.5	17.5	16.5	15.7	15.3	15.2	14.7	14.5	14.1	14.3	15.57
	Mean	17.96	..	18.06	19.89	21.37	22.43	23.46	24.51	24.94	24.84	24.21	23.46	22.62	21.75	20.83	20.10	19.65	19.43	19.06	18.42	18.08	20.06
IX 1889	1	13.8	..	13.2	..	..	..	17.5	..	..	..	18.0	..	..	..	15.8	..	..	..	14.9	..	..	15.45
	2	15.3	..	14.8	..	..	..	18.5	..	..	..	20.9	..	..	..	19.1	..	..	..	19.2	..	..	17.97
	3	18.7	..	18.1	..	..	..	19.7	..	..	..	21.1	..	..	..	19.1	..	..	..	16.1	..	..	18.80
	4	15.2	..	14.3	..	..	..	18.3	..	..	..	18.6	..	..	..	15.7	..	..	..	15.1	..	..	16.20
	5	14.9	..	14.4	..	..	..	17.6	..	..	..	19.1	..	..	..	17.2	..	..	..	16.9	..	..	16.68
	6	17.1	..	18.5	..	..	..	23.8	..	..	..	22.9	..	..	..	21.4	..	..	..	19.1	..	..	20.47
	7	17.9	..	17.8	..	..	..	18.9	..	..	..	17.1	..	..	..	16.9	..	..	..	16.1	..	..	17.45
General Mean		17.62	..	17.06	19.89	21.37	22.43	22.65	24.51	24.94	24.84	23.37	23.46	22.62	21.75	20.27	20.10	19.65	19.43	18.64	18.42	18.08	20.31
										KUROSAWA 832m.													
VIII 1894	1	14.1	12.2	12.6	..	19.3	..	25.7	..	27.1	..	27.8	..	27.4	..	23.5	..	19.2	..	18.8	..	18.6	20.52
	2	17.5	16.8	17.7	..	20.8	..	25.2	..	23.2	..	22.6	..	19.6	..	18.9	..	18.1	..	17.8	..	17.4	19.63
	3	17.1	17.4	17.5	..	20.4	..	23.9	..	26.2	..	26.7	..	26.0	..	23.9	..	21.2	..	20.0	..	20.8	21.76
	4	20.8	20.6	19.4	..	22.1	..	21.7	..	23.2	..	25.5	..	22.8	..	21.3	..	19.2	..	19.0	..	18.4	21.19
	5	18.0	17.8	18.0	..	19.9	..	22.1	..	24.7	..	25.0	..	22.6	..	21.6	..	20.6	..	19.9	..	19.2	20.78
	6	19.4	19.2	19.0	..	20.6	..	24.1	..	26.1	..	25.3	..	22.8	..	23.2	..	20.6	..	20.0	..	19.4	21.64
	7	19.3	19.0	19.2	..	22.1	..	25.7	..	26.6	..	26.1	..	23.7	..	23.0	..	21.2	..	19.6	..	18.2	21.97
	8	17.8	16.3	16.2	..	20.9	..	24.7	..	28.4	..	29.2	..	30.4	..	25.6	..	19.9	..	18.6	..	17.8	22.23
	9	17.3	16.8	16.5	..	21.2	..	27.0	..	30.2	..	28.4	..	29.7	..	25.6	..	21.4	..	19.2	..	18.4	22.64
	10	17.8	17.8	17.8	..	22.7	..	24.9	..	27.0	..	28.4	..	26.2	..	24.2	..	20.6	..	19.0	..	18.8	22.10
	11	18.8	17.9	17.8	..	21.2	..	25.2	..	25.4	..	27.0	..	27.1	..	24.4	..	20.2	..	19.3	..	18.6	21.91
	12	17.8	16.8	17.0	..	22.2	..	26.7	..	28.8	..	29.5	..	28.0	..	24.2	..	20.0	..	19.1	..	18.9	22.42
	13	18.4	17.5	18.0	..	21.2	..	27.1	..	29.2	..	28.6	..	28.4	..	25.0	..	20.2	..	18.3	..	18.2	22.51
	14	16.4	15.0	15.5	..	21.9	..	26.4	..	30.1	..	29.2	..	30.1	..	26.2	..	23.6	..	22.2	..	20.0	23.05
	15	20.6	20.4	20.0	..	24.2	..	28.0	..	29.0	..	29.4	..	29.2	..	25.6	..	20.3	..	19.3	..	17.3	23.61

TABLE II. AIR TEMPERATURE.

Month & Year	Day	2h am	4h am	6h am	7h am	8h am	9h am	10h am	11h am	Noon	1h pm	2h pm	3h pm	4h pm	5h pm	6h pm	8h pm	10h pm	Mid-night	Mean
ONTAKE 3062m.																				
VIII 1891	16	7.0	7.6	6.2	..	7.9	..	8.4	..	9.2	..	9.0	..	8.4	..	8.5	7.7	8.4	9.4	8.14
	17	9.3	9.8	9.2	..	9.7	..	8.7	..	9.7	..	9.2	..	9.5	..	9.0	9.4	9.7	9.1	9.22
	18	9.0	9.3	9.0	..	9.1	..	9.3	..	9.8	..	9.2	..	9.4	..	9.1	3.8	4.5	3.1	7.88
	19	2.1	1.3	2.4	..	6.0	..	8.2	..	11.1	..	10.7	..	8.4	..	6.4	5.3	5.0	3.8	5.89
	20	3.2	3.3	5.1	..	9.0	..	12.8	..	12.1	..	12.5	..	8.9	..	5.4	5.0	4.8	3.8	7.16
	21	2.5	4.7	5.4	..	4.8	..	7.2	..	8.0	..	8.9	..	7.1	..	7.9	7.6	7.9	8.4	6.71
	22	8.6	6.8	5.4	..	6.4	..	8.2	..	10.0	..	8.6	..	7.3	..	7.2	7.4	7.6	7.5	7.58
	23	7.0	6.2	7.2	..	7.8	..	9.0	..	8.7	..	8.1	..	6.9	..	5.2	5.3	5.6	4.4	6.78
	24	4.6	4.4	4.8	..	6.6	..	8.8	..	11.6	..	10.4	..	9.1	..	5.4	4.8	4.8	5.0	6.69
	25	4.0	3.4	3.7	..	5.2	..	8.1	..	10.7	..	13.4	..	10.2	..	6.0	4.8	2.9	3.8	6.43
	26	3.8	2.9	4.1	..	9.9	..	13.2	..	13.4	..	12.3	..	11.2	..	6.2	5.0	4.8	4.1	7.57
	27	4.4	5.1	5.9	..	11.4	..	12.5	..	14.8	..	10.2	..	8.3	..	6.3	5.9	3.8	3.8	7.67
	28	3.9	3.8	4.7	..	5.2	..	11.6	..	10.1	..	9.4	..	8.6	..	6.7	5.6	5.4	4.6	6.63
	29	4.2	5.0	4.7	..	5.4	..	8.1	..	5.0	..	11.6	..	10.4	..	7.2	6.3	6.9	6.4	6.77
	30	5.7	5.4	7.8	..	10.8	..	18.2	..	18.3	..	16.3	..	13.5	..	8.3	7.0	7.2	7.0	10.46
	31	7.1	6.6	8.0	..	13.3	..	17.6	..	15.7	..	13.0	..	10.7	..	9.4	7.5	9.2	8.9	10.58
	Mean	6.64	6.40	6.86	..	8.75	..	11.07	..	12.13	..	11.97	..	10.40	..	8.42	7.06	7.07	6.86	8.61
IX 1891	1	6.2	6.2	7.5	..	9.3	..	15.8	..	19.2	..	16.5	..	14.1	..	9.5	7.9	8.3	8.4	10.74
	2	8.4	8.2	7.9	..	8.9	..	11.9	..	16.7	..	16.0	..	12.9	..	10.3	9.0	9.2	9.3	10.72
	3	8.6	8.9	8.9	..	11.7	..	18.5	..	15.7	..	15.0	..	15.3	..	12.3	9.6	9.7	8.9	11.92
	4	9.2	9.2	9.0	..	12.9	..	17.2	..	18.0	..	16.1	..	11.7	..	10.5	8.8	8.6	7.2	11.53
	5	7.4	6.8	6.0	..	7.1	..	9.5	..	17.9	..	19.0	..	14.7	..	9.7	7.8	7.4	8.4	10.14
	6	8.0	7.7	8.4	..	14.6	..	18.2	..	17.9	..	17.7	..	16.0	..	9.9	8.0	7.1	7.4	11.74
	7	7.5	7.2	7.0	..	8.7	..	16.1	..	13.2	..	13.0	..	10.0	..	8.7	7.4	6.5	7.0	9.36
	8	6.5	6.6	5.8	..	7.5	..	7.4	..	7.2	..	7.6	..	5.7	..	6.7	6.0	6.8	8.1	6.82
	9	7.9	8.2	8.8	..	9.2	..	9.6	..	9.9	..	9.8	..	9.4	..	8.2	8.8	9.0	7.9	8.83
	10	7.6	7.7	6.4	..	5.8	..	7.0	..	7.4	..	8.2	..	9.3	..	9.9	10.1	10.3	7.8	8.12
	11	8.9	7.6	9.4	..	8.9	..	9.9	..	10.6	..	9.8	..	9.4	..	8.2	6.8	8.2	7.0	8.65
	12	6.8	6.2	6.3	..	8.2	..	12.8	..	12.4	..	12.6	..	10.0	..	8.5	8.6	9.2	9.0	9.22
General Mean		6.93	6.72	7.07	..	8.93	..	11.52	..	12.60	..	12.38	..	10.72	..	8.47	7.41	7.43	7.19	8.93
GOZAISHODAKE 1200m.																				
IX 1888	4	19.0	18.4	18.7	..	18.4	..	18.5	..	18.9	..	18.4	..	18.1	..	18.2	17.0	16.8	16.8	18.0
	5	16.4	15.4	14.4	..	16.5	..	15.6	..	17.1	..	16.5	..	16.5	..	13.0	12.0	11.4	11.0	14.5
	6	10.9	10.3	10.2	..	11.4	..	14.8	..	18.4	..	17.4	..	17.0	..	12.7	12.7	12.0	12.4	13.3
	7	12.6	11.8	12.7	..	13.9	..	16.6	..	19.8	..	19.7	..	18.2	..	15.4	15.3	15.0	15.5	15.6
	8	14.6	14.7	15.1	..	15.6	..	16.2	..	16.4	..	16.6	..	16.1	..	16.4	15.6	15.5	15.3	15.7
	9	14.0	13.5	13.6	..	15.2	..	18.0	..	15.6	..	16.3	..	16.6	..	15.9	14.8	15.3	15.3	15.3
	10	15.6	14.6	15.1	..	16.0	..	16.4	..	17.0	..	17.8	..	17.4	..	17.1	16.4	14.6	14.9	16.0
	11	14.0	14.0	14.0	..	14.0	..	14.6	..	14.8	..	15.0	..	15.2	..	15.4	15.5	15.5	16.2	14.9
	12	18.0	17.5	16.2	..	17.5	..	17.3	..	17.6	..	17.3	..	15.6	..	14.6	14.0	13.6	12.6	16.0
	13	11.9	9.5	10.4	..	14.4	..	16.2	..	18.0	..	10.1	..	11.6	..	9.1	8.7	8.7	8.7	11.4
	14	9.4	9.3	8.8	..	10.0	..	13.0	..	13.4	..	15.2	..	11.6	..	9.9	10.1	9.7	9.5	10.9
	15	11.4	12.0	11.7	..	13.6	..	15.9	..	18.9	..	17.9	..	17.4	..	13.4	13.1	13.9	12.3	14.3
	16	11.8	11.4	12.2	..	12.7	..	14.0	..	15.2	..	16.7	..	16.0	..	13.9	13.7	13.1	13.5	13.7
	17	12.5	12.2	11.6	..	12.4	..	15.0	..	15.0	..	15.4	..	14.9	..	11.8	11.3	11.8	11.6	12.9
	18	11.2	11.2	11.6	..	12.2	..	15.2	..	16.7	..	16.6	..	13.7	..	10.8	9.9	9.1	10.8	13.4
	19	10.9	11.2	11.5	..	13.6	..	14.4	..	15.6	..	14.9	..	13.6	..	13.6	13.6	14.2	14.2	13.5
	20	14.2	14.8	14.0	..	14.4	..	14.9	..	15.4	..	15.4	..	15.6	..	15.2	15.8	15.6	15.2	15.1
	21	14.9	14.6	14.2	..	14.9	..	15.2	..	16.6	..	16.5	..	16.8	..	14.2	14.2	14.1	13.6	14.9
	22	14.1	15.2	13.3	..	13.7	..	15.1	..	15.8	..	14.8	..	14.9	..	14.4	14.4	13.8	12.6	14.3
	23	12.4	12.8	12.3	..	13.2	..	15.4	..	17.4	..	17.0	..	16.7	..	13.1	12.6	12.9	12.4	14.0
	24	12.1	12.2	12.9	..	13.4	..	15.7	..	18.1	..	14.0	..	12.1	..	10.1	8.7	8.5	8.2	12.1
	25	8.0	10.1	11.4	..	14.1	..	14.0	..	14.0	..	13.4	..	13.0	..	12.3	12.6	12.9	12.4	12.3
	26	11.8	11.8	12.0	..	13.3	..	17.0	..	17.1	..	16.4	..	14.9	..	13.6	12.8	12.9	12.9	13.8
	27	12.4	11.8	11.9	..	13.6	..	15.3	..	16.8	..	18.9	..	14.8	..	13.8	13.2	12.6	12.0	13.9
	28	11.7	11.4	11.4	..	13.9	..	15.8	..	15.8	..	16.4	..	14.3	..	12.3	11.8	11.2	11.0	13.1
	29	10.1	9.6	10.3	..	12.8	..	15.9	..	14.3	..	13.5	..	13.7	..	10.8	9.1	9.1	8.7	11.5
	30	8.3	7.8	7.6	..	10.6	..	11.5	..	15.0	..	13.8	..	11.2	..	9.7	9.3	9.2	8.5	10.8
X 1888	1	8.4	8.1	8.3	..	8.9	..	9.3	..	11.8	..	13.3	..	9.3	..	8.3	7.0	6.2	6.2	8.7
	2	5.6	5.2	4.0	..	5.6	..	8.2	..	9.5	..	9.9	..	8.3	..	5.1	5.6	4.8	4.8	6.4
	3	4.7	4.2	4.0	..	6.4	..	8.5	..	9.7	..	11.4	..	9.1	..	5.7	5.2	5.4	..	6.7
	Mean	12.1	11.9	11.8	..	13.1	..	14.8	..	15.8	..	15.5	..	14.4	..	12.7	12.2	12.0	11.6	13.6

Month & Year	Day	2h am	4h am	6h am	7h am	8h am	9h am	10h am	11h am	Noon	1h pm	2h pm	3h pm	4h pm	5h pm	6h pm	7h pm	8h pm	9h pm	10h pm	11h pm	Mid-night	Mean
										KUROSAWA 832m.													
VIII 1891	16	16.4	14.9	15.4	..	20.2	..	26.8	..	28.2	..	27.9	..	25.0	..	22.6	..	20.2	..	20.7	..	20.0	21.52
	17	19.2	20.3	21.2	..	20.7	..	20.3	..	23.2	..	24.6	..	22.0	..	20.3	..	20.2	..	20.0	..	19.2	20.93
	18	19.4	19.3	19.5	..	20.3	..	22.5	..	26.0	..	21.1	..	20.8	..	19.7	..	17.0	..	16.2	..	15.6	19.78
	19	15.0	13.2	12.8	..	16.8	..	23.2	..	25.4	..	24.9	..	25.2	..	21.5	..	26.2	..	14.6	..	12.4	18.43
	20	11.5	10.4	11.2	..	15.1	..	21.7	..	24.0	..	24.4	..	24.2	..	20.2	..	15.6	..	13.8	..	13.8	17.16
	21	13.7	13.8	14.2	..	14.2	..	15.6	..	16.3	..	18.2	..	19.0	..	18.4	..	19.0	..	18.7	..	18.7	16.65
	22	18.5	18.2	17.6	..	19.6	..	20.6	..	21.7	..	21.2	..	21.6	..	18.9	..	18.2	..	17.9	..	17.6	19.30
	23	17.4	17.1	17.4	..	18.5	..	19.4	..	19.8	..	18.9	..	19.2	..	18.7	..	17.4	..	17.0	..	16.0	18.07
	24	15.7	15.0	14.9	..	17.2	..	20.3	..	22.2	..	22.5	..	22.0	..	19.7	..	18.2	..	17.5	..	16.9	18.50
	25	16.0	15.7	16.2	..	19.7	..	24.4	..	25.6	..	25.9	..	23.5	..	20.8	..	16.8	..	15.6	..	15.0	19.60
	26	14.1	13.9	13.5	..	14.9	..	21.8	..	25.4	..	26.5	..	26.4	..	18.2	..	14.8	..	12.9	..	11.6	17.83
	27	10.0	9.4	9.0	..	16.4	..	22.2	..	26.6	..	26.4	..	25.1	..	20.4	..	17.0	..	15.3	..	14.0	17.72
	28	14.2	12.4	12.1	..	16.5	..	21.8	..	25.6	..	26.7	..	25.9	..	20.8	..	16.8	..	15.1	..	14.2	18.51
	29	14.0	13.6	13.8	..	16.4	..	21.6	..	25.8	..	28.3	..	27.2	..	21.1	..	17.6	..	15.2	..	13.0	18.97
	30	11.8	10.0	10.4	..	14.6	..	22.3	..	27.2	..	28.4	..	25.0	..	20.0	..	16.6	..	14.1	..	12.6	17.75
	31	11.2	10.8	11.0	..	17.0	..	23.7	..	27.8	..	28.2	..	27.2	..	22.3	..	17.9	..	17.2	..	17.4	19.31
	Mean	16.43	15.79	15.88	..	19.33	..	23.44	..	25.71	..	25.90	..	24.95	..	21.96	..	18.93	..	17.80	..	17.09	20.26
IX 1891	1	15.6	14.6	14.3	..	20.3	..	26.0	..	28.4	..	28.0	..	27.8	..	23.6	..	19.6	..	18.6	..	18.0	21.23
	2	17.2	16.5	16.4	..	21.0	..	26.2	..	28.5	..	26.2	..	24.2	..	21.6	..	19.3	..	18.6	..	17.8	21.12
	3	17.6	16.8	16.8	..	20.4	..	26.4	..	31.2	..	29.5	..	29.4	..	23.0	..	20.9	..	19.2	..	19.6	22.57
	4	19.0	18.0	17.6	..	21.2	..	25.8	..	28.5	..	28.7	..	27.8	..	22.7	..	19.6	..	17.8	..	17.0	21.97
	5	16.1	16.0	15.8	..	19.9	..	26.7	..	28.6	..	29.8	..	26.2	..	22.5	..	18.4	..	17.5	..	15.4	21.07
	6	15.8	15.8	14.7	..	18.0	..	24.3	..	28.4	..	29.4	..	27.9	..	22.6	..	18.4	..	17.7	..	14.9	20.66
	7	14.3	13.2	13.4	..	17.8	..	23.8	..	25.6	..	26.2	..	24.2	..	21.4	..	19.6	..	17.6	..	17.6	19.56
	8	17.2	16.9	16.6	..	20.2	..	27.9	..	24.6	..	24.4	..	22.4	..	20.2	..	19.1	..	18.4	..	18.0	20.07
	9	18.2	18.0	18.2	..	18.6	..	19.6	..	23.1	..	22.3	..	23.2	..	20.8	..	20.1	..	19.4	..	19.1	20.05
	10	18.8	18.4	18.4	..	21.2	..	25.5	..	26.0	..	26.0	..	26.3	..	22.8	..	20.6	..	20.4	..	19.0	21.95
	11	17.8	16.6	17.4	..	21.4	..	27.8	..	27.6	..	29.4	..	28.6	..	22.4	..	20.2	..	18.5	..	17.0	22.07
	12	17.0	15.9	15.0	..	20.4	..	26.8	..	28.9	..	30.3	..	29.4	..	21.8	..	20.2	..	19.6	..	19.3	22.05
General Mean		16.68	15.96	15.98	..	19.52	..	23.62	..	26.20	..	25.85	..	25.37	..	22.08	..	19.13	..	18.03	..	17.23	20.52
										YOKKAICHI 4m.													
IX 1888	4	24.4	..	25.6	..	..	..	31.1	..	..	..	30.1	..	..	..	26.0	..	..	..	23.6	..	..	26.8
	5	23.4	..	23.0	..	..	..	25.7	..	..	..	27.6	..	..	..	23.4	..	..	..	20.0	..	..	23.8
	6	16.8	..	16.4	..	..	..	25.2	..	..	..	27.5	..	..	..	24.0	..	..	..	19.0	..	..	21.5
	7	19.4	..	18.8	..	..	..	25.2	..	..	..	27.9	..	..	..	25.3	..	..	..	21.0	..	..	22.9
	8	20.6	..	19.7	..	..	..	22.9	..	..	..	24.5	..	..	..	23.4	..	..	..	22.3	..	..	22.2
	9	21.5	..	20.3	..	..	..	24.2	..	..	..	23.4	..	..	..	23.8	..	..	..	21.6	..	..	22.5
	10	21.1	..	21.6	..	..	..	21.4	..	..	..	22.7	..	..	..	22.2	..	..	..	19.0	..	..	21.3
	11	19.2	..	20.0	..	..	..	20.9	..	..	..	22.0	..	..	..	24.2	..	..	..	23.5	..	..	21.6
	12	23.2	..	23.8	..	..	..	24.2	..	..	..	25.6	..	..	..	23.5	..	..	..	21.4	..	..	23.6
	13	18.7	..	16.9	..	..	..	24.8	..	..	..	21.4	..	..	..	23.8	..	..	..	16.5	..	..	20.3
	14	16.4	..	15.7	..	..	..	22.5	..	..	..	21.9	..	..	..	19.5	..	..	..	18.8	..	..	19.6
	15	15.9	..	15.2	..	..	..	22.4	..	..	..	25.3	..	..	..	23.5	..	..	..	20.9	..	..	20.5
	16	21.3	..	19.0	..	..	..	21.0	..	..	..	22.2	..	..	..	21.2	..	..	..	19.0	..	..	20.6
	17	18.0	..	17.2	..	..	..	24.9	..	..	..	25.8	..	..	..	21.2	..	..	..	17.9	..	..	21.0
	18	16.2	..	16.7	..	..	..	24.0	..	..	..	26.7	..	..	..	20.7	..	..	..	16.8	..	..	20.2
	19	14.6	..	14.7	..	..	..	22.8	..	..	..	24.5	..	..	..	23.2	..	..	..	22.8	..	..	20.4
	20	20.9	..	21.1	..	..	..	21.4	..	..	..	23.4	..	..	..	23.2	..	..	..	23.4	..	..	22.2
	21	21.7	..	20.5	..	..	..	20.5	..	..	..	22.4	..	..	..	22.2	..	..	..	21.7	..	..	21.5
	22	21.3	..	20.0	..	..	..	23.9	..	..	..	21.9	..	..	..	21.2	..	..	..	20.4	..	..	21.4
	23	19.1	..	18.0	..	..	..	25.9	..	..	..	27.6	..	..	..	22.2	..	..	..	18.5	..	..	21.9
	24	15.9	..	14.5	..	..	..	22.2	..	..	..	24.8	..	..	..	20.0	..	..	..	17.2	..	..	19.1
	25	15.0	..	14.8	..	..	..	20.2	..	..	..	20.8	..	..	..	20.4	..	..	..	19.4	..	..	18.4
	26	18.9	..	17.6	..	..	..	23.5	..	..	..	25.1	..	..	..	21.2	..	..	..	19.1	..	..	20.9
	27	19.2	..	17.5	..	..	..	22.0	..	..	..	24.2	..	..	..	22.2	..	..	..	18.4	..	..	20.6
	28	14.5	..	12.6	..	..	..	22.0	..	..	..	25.5	..	..	..	21.2	..	..	..	18.8	..	..	19.3
	29	17.7	..	14.7	..	..	..	22.8	..	..	..	25.2	..	..	..	19.4	..	..	..	15.8	..	..	19.3
	30	14.8	..	14.7	..	..	..	22.2	..	..	..	24.3	..	..	..	19.9	..	..	..	16.8	..	..	18.8
X 1888	1	16.0	..	15.9	..	..	..	19.8	..	..	..	21.6	..	..	..	16.4	..	..	..	15.3	..	..	17.5
	2	15.1	..	13.0	..	..	..	19.5	..	..	..	21.4	..	..	..	14.9	..	..	..	13.4	..	..	16.2
	3	11.5	..	9.3	..	..	..	19.6	..	..	..	21.9	..	..	..	15.0	..	..	..	10.3	..	..	14.6
	Mean	18.4	..	17.6	..	..	..	22.9	..	..	..	24.5	..	..	..	21.6	..	..	..	19.1	..	..	20.7

TABLE II. AIR TEMPERATURE.

Month & Year	Day	2h a.m.	6h a.m.	10h a.m.	2h p.m.	6h p.m.	10h p.m.	Mean
HIGASHI HOBEN 736m.								
		°	°	°	°	°	°	°
VIII 1889	1	21.1	21.1	23.9	26.3	23.9	21.1	22.90
	2	21.7	21.2	25.6	24.8	24.5	22.3	23.35
	3	23.4	22.8	23.4	26.9	23.0	23.3	23.80
	4	23.6	22.0	23.7	23.0	22.6	23.8	23.12
	5	23.9	22.3	26.7	28.2	24.8	22.8	24.78
	6	21.6	21.4	23.9	26.3	24.7	22.4	23.38
	7	21.1	21.1	24.7	26.5	24.2	21.7	23.22
	8	20.7	20.6	24.8	26.7	24.2	22.2	23.20
	9	21.6	21.0	24.2	27.8	22.7	22.3	23.27
	10	23.0	22.6	24.9	24.8	22.5	23.5	23.55
	11	23.5	21.7	25.1	26.9	22.4	21.8	23.57
	12	20.5	20.1	22.2	26.0	22.4	22.0	22.20
	13	21.7	21.9	23.9	27.0	22.0	20.1	22.77
	14	20.3	20.9	22.1	22.7	19.2	19.5	20.78
	15	18.9	19.1	20.8	21.5	19.6	19.9	19.97
	16	19.0	19.1	19.3	22.5	20.9	20.3	20.18
	17	20.6	20.6	22.2	23.7	23.2	21.8	22.02
	18	21.8	21.5	25.6	25.0	22.9	21.5	23.05
	19	21.8	21.1	22.4	20.2	19.6	19.8	20.82
	20	20.0	20.2	20.6	21.1	20.8	20.4	20.52
	21	19.7	20.3	23.5	25.9	24.0	22.9	22.72
	22	21.7	21.3	22.8	25.9	22.6	21.5	22.63
	23	21.9	21.6	23.0	24.3	22.5	22.1	22.57
	24	21.8	21.5	21.6	22.1	21.4	20.9	21.55
	25	21.2	21.3	21.7	21.9	21.3	20.8	21.37
	26	20.0	18.8	18.7	18.5	17.9	17.3	18.53
	27	16.9	16.8	18.2	19.7	17.9	17.2	17.78
	28	16.8	16.7	19.3	19.6	17.7	17.5	17.93
	29	16.0	15.5	17.8	18.8	17.6	17.1	17.05
	30	18.3	16.5	20.3	18.5	18.5	18.5	18.43
	31	18.4	17.4	18.2	18.8	18.4	17.6	18.13
	Mean	20.73	20.32	22.42	23.59	21.61	20.83	21.58
IX 1889	1	17.3	17.5	19.1	21.5	19.9	19.8	19.18
	2	19.6	19.8	19.9	18.6	17.6	17.3	18.80
	3	16.8	16.0	22.2	22.0	18.9	19.9	19.30
	4	19.4	19.3	20.3	22.4	20.1	18.6	20.02
	5	18.9	19.4	19.7	22.6	21.1	20.3	20.30
	6	20.3	19.1	19.1	19.1	17.9	17.4	18.45
	7	16.9	17.0	19.7	20.9	17.7	17.3	18.25
	8	16.9	16.6	20.9	21.3	18.2	16.8	18.45
	9	16.9	16.6	18.0	20.8	17.1	15.8	17.43
	10	15.9	15.7	17.8	19.6	17.2	16.1	17.05
	11	15.7	15.7	17.0	15.5	14.9	14.8	15.60
	12	15.3	14.9	15.2	17.4	15.7	14.8	15.55
	13	14.5	14.1	16.7	17.6	15.5	15.4	15.63
	14	13.8	13.7	16.6	17.9	15.1	14.5	15.27
	15	14.0	13.8	16.9	16.3	15.4	14.3	15.12
	16	13.7	13.0	16.5	17.3	15.1	14.2	14.97
	17	14.0	13.4	17.3	18.2	14.9	14.2	15.33
	18	15.2	14.9	20.2	19.4	16.2	15.8	16.95
	19	15.7	15.5	18.9	18.9	16.6	15.5	16.85
	20	15.7	16.7	17.5	18.2	18.4	19.1	17.60
	21	18.8	19.0	18.7	18.4	16.9	15.9	17.95
	22	15.6	14.2	14.6	17.0	16.3	16.4	15.83
	23	14.6	13.9	16.3	16.5	14.3	13.5	14.85
	24	13.3	14.1	15.6	17.6	14.1	13.7	14.73
	25	13.3	13.5	14.8	16.1	14.3	13.4	14.13
	26	12.7	13.4	17.3	18.0	15.6	14.5	15.25
	27	15.0	14.6	13.9	16.4	15.1	14.5	14.92

TABLE IIa. AIR TEMPERATURE.

Month & Year	Day	2h a.m.	6h a.m.	10h a.m.	2h p.m.	6h p.m.	10h p.m.	Mean
YAMAGUCHI 35m.								
		°	°	°	°	°	°	°
VIII 1889	1	24.2	23.0	29.7	31.8	29.4	24.6	27.12
	2	24.3	24.6	30.5	31.1	32.2	25.7	28.07
	3	24.0	22.8	31.1	33.8	30.4	25.2	27.88
	4	22.4	21.4	30.2	29.8	27.6	23.9	25.88
	5	22.0	20.7	29.7	32.8	30.9	24.3	26.73
	6	23.2	23.9	29.8	31.8	31.5	26.3	27.75
	7	24.5	24.7	29.6	32.3	31.8	23.9	27.80
	8	22.5	23.1	29.4	33.1	30.7	22.8	26.93
	9	21.2	20.4	29.6	33.3	29.5	24.7	26.45
	10	22.8	21.8	29.4	33.4	29.2	24.0	26.77
	11	23.5	22.5	30.3	34.1	30.2	23.6	27.37
	12	21.5	21.2	28.8	33.4	30.1	24.2	26.53
	13	21.8	21.0	29.8	32.9	29.3	23.8	26.43
	14	22.7	24.9	28.4	29.9	25.6	25.4	26.15
	15	24.6	23.7	28.2	29.0	25.8	26.1	26.23
	16	24.2	24.2	26.9	29.9	27.9	26.4	26.58
	17	25.7	24.3	30.2	30.5	30.8	24.4	27.48
	18	22.8	21.8	31.4	32.0	29.6	27.6	27.63
	19	27.0	27.5	31.6	27.7	26.3	25.8	27.65
	20	24.9	24.0	26.3	29.0	28.3	23.9	26.07
	21	22.4	21.5	29.5	31.4	30.4	24.2	26.57
	22	23.1	23.5	29.7	32.3	28.8	25.3	27.12
	23	26.1	25.6	30.1	31.5	28.7	27.6	28.27
	24	27.1	26.7	29.3	30.2	27.9	26.5	27.95
	25	25.9	25.9	27.6	26.4	25.6	24.9	26.05
	26	24.8	23.9	26.6	27.6	24.3	23.0	25.03
	27	22.3	22.0	26.5	28.0	24.8	22.6	24.37
	28	22.5	21.7	27.3	28.3	24.6	23.5	24.65
	29	21.3	18.5	26.4	25.4	24.3	22.5	23.07
	30	21.2	20.9	27.2	25.7	25.0	20.6	23.43
	31	20.2	21.5	23.2	23.2	23.3	21.5	22.15
	Mean	23.44	23.01	28.82	30.39	28.22	24.48	26.39
IX 1889	1	21.4	21.2	25.1	27.9	26.9	24.7	24.38
	2	24.6	24.0	22.3	26.6	23.8	21.4	23.95
	3	20.1	19.2	26.3	29.6	25.2	21.1	23.58
	4	20.2	19.2	26.1	28.7	26.9	21.8	23.82
	5	21.5	22.5	26.2	28.7	27.1	24.9	25.15
	6	23.9	23.9	28.4	26.3	23.9	21.4	24.30
	7	19.3	18.6	26.3	28.5	24.2	20.9	22.82
	8	17.8	16.8	26.4	27.8	24.9	21.9	22.60
	9	19.8	17.6	25.5	28.2	23.7	21.0	22.63
	10	20.2	17.5	25.4	26.8	24.1	20.7	22.45
	11	20.7	20.8	24.1	23.7	20.7	20.1	21.68
	12	20.1	18.6	23.5	25.2	22.4	19.8	21.60
	13	15.9	14.2	24.2	25.0	21.8	17.4	19.75
	14	14.2	13.2	23.8	26.6	21.4	16.4	19.27
	15	14.1	12.3	23.3	22.9	21.0	15.6	18.20
	16	13.4	10.7	23.4	25.3	21.3	18.1	18.70
	17	14.2	12.7	23.3	26.8	21.1	17.6	19.28
	18	14.4	12.0	23.7	27.6	22.3	17.1	19.52
	19	14.8	15.2	24.9	26.1	20.8	19.7	20.25
	20	19.5	20.7	22.9	23.2	23.3	23.0	22.10
	21	22.3	22.2	23.3	24.0	22.7	19.6	22.35
	22	17.6	18.1	21.7	24.5	20.9	16.4	19.87
	23	16.3	15.6	21.0	22.0	19.7	15.6	18.37
	24	13.5	13.9	18.2	24.2	20.0	14.2	17.33
	25	13.7	13.5	18.5	22.2	20.1	16.3	17.38
	26	14.0	12.1	22.0	24.8	20.9	18.3	18.68
	27	18.3	18.4	18.8	20.5	19.5	18.5	19.00

TABLE II. AIR TEMPERATURE.

Month & Year	Day	2h am	6h am	10h am	2h pm	6h pm	10h pm	Mean
HIGASHI HOBEN 736m.								
IX 1889		°	°	°	°	°	°	°
	28	14.0	14.4	15.7	16.1	14.4	13.3	14.65
	29	12.6	12.3	13.2	15.4	12.7	12.5	13.12
	30	12.3	10.1	12.0	12.7	10.2	10.2	11.27
	Mean	15.62	15.41	17.48	18.28	16.20	15.61	16.43
X 1889	1	10.1	9.9	11.9	13.5	10.7	10.4	11.08
	2	11.2	10.6	13.3	15.5	12.7	12.1	12.58
	3	12.9	11.8	14.6	17.5	13.9	13.8	14.08
	4	13.4	12.5	15.8	18.8	14.1	13.6	14.70
	5	15.5	15.2	17.4	17.9	15.2	15.7	16.15
	6	14.6	12.9	14.8	14.4	13.8	13.6	14.02
	7	13.3	12.9	14.9	16.8	12.6	12.1	13.63
	8	11.8	11.7	17.2	15.5	12.7	13.1	13.67
	9	12.5	12.1	13.4	15.3	13.0	13.7	13.33
	10	13.0	12.5	15.5	17.2	13.1	13.1	14.07
	11	13.3	13.4	17.0	17.9	14.6	14.1	15.05
	12	13.7	12.6	15.9	20.1	15.3	13.1	15.12
	13	13.0	12.5	13.7	16.5	12.7	12.4	13.47
	14	10.4	10.2	12.9	14.6	13.0	11.7	12.13
	15	12.7	13.2	14.7	14.7	14.9	14.8	14.17
	16	14.1	12.5	13.0	13.9	13.0	11.8	13.05
	17	11.4	11.1	14.0	16.8	13.0	13.6	13.32
	18	13.6	12.5	16.3	17.2	13.0	12.9	14.25
	19	12.7	14.1	17.9	15.2	13.9	13.0	14.47
	20	12.1	12.4	15.6	15.8	14.4	15.4	14.28
	21	17.7	17.1	14.3	11.0	8.7	7.8	12.77
	22	7.0	6.5	7.7	8.0	6.4	6.8	7.07
	23	6.8	7.0	9.2	9.9	8.2	8.0	8.18
	24	8.6	8.5	11.1	12.7	9.4	8.9	9.87
	25	8.3	8.4	11.8	15.3	10.0	9.5	10.55
	26	9.2	9.3	11.9	14.7	11.1	10.9	11.18
	27	10.9	10.9	14.8	17.5	13.2	13.3	13.43
	28	13.0	12.8	13.4	13.4	13.4	12.7	13.12
	29	12.4	7.0	3.1	2.0	1.4	1.8	4.62
	30	1.6	1.0	3.0	7.6	3.7	4.2	3.52
	31	5.4	5.8	11.0	9.9	8.4	8.3	8.13
	Mean	11.49	11.00	13.29	14.40	11.72	11.49	12.23

TABLE IIa. AIR TEMPERATURE.

Month & Year	Day	2h am	6h am	10h am	2h pm	6h pm	10h pm	Mean
YAMAGUCHI 35m.								
IX 1889		°	°	°	°	°	°	°
	28	17.3	16.5	20.1	24.0	20.4	18.6	19.45
	29	17.4	16.4	20.4	22.2	17.6	16.0	18.33
	30	15.7	16.2	19.2	20.5	15.3	12.4	16.55
	Mean	17.87	17.12	23.24	25.35	22.10	18.09	20.49
X 1889	1	10.0	7.6	19.3	21.7	16.6	12.2	14.57
	2	11.6	10.4	16.3	21.2	17.9	14.3	15.28
	3	14.2	11.8	21.2	24.7	17.0	13.2	17.02
	4	12.9	11.2	22.2	25.6	19.2	12.7	17.30
	5	11.4	11.3	20.9	24.7	19.9	19.2	17.90
	6	18.7	17.3	19.6	21.3	19.5	18.4	19.13
	7	14.8	12.2	21.6	23.7	19.1	16.0	17.90
	8	15.2	11.8	21.2	23.0	18.9	15.5	17.60
	9	15.3	16.3	20.1	22.4	18.4	13.7	17.70
	10	11.8	10.7	20.5	25.2	19.1	13.3	16.77
	11	10.6	10.0	19.4	24.1	17.6	16.3	16.17
	12	15.0	14.6	20.9	24.3	19.6	16.5	18.48
	13	11.8	9.8	21.2	24.2	16.7	11.6	15.88
	14	8.9	8.2	17.3	21.1	17.3	15.3	14.68
	15	15.6	15.6	18.7	20.7	18.9	18.0	17.92
	16	18.2	18.5	20.9	21.5	18.3	14.1	18.65
	17	10.3	9.5	19.2	22.9	17.2	15.2	15.72
	18	13.5	13.1	21.8	22.2	17.9	13.1	16.93
	19	11.0	10.7	18.7	21.1	17.6	13.3	15.40
	20	12.7	12.4	22.9	22.3	19.5	20.5	18.38
	21	22.0	20.7	20.0	18.1	15.2	13.5	18.25
	22	11.8	10.1	15.5	15.3	11.7	8.7	12.18
	23	6.0	4.2	14.9	17.9	11.6	9.3	10.65
	24	9.3	9.5	13.4	18.2	14.3	11.3	12.67
	25	7.2	6.3	16.7	20.3	14.9	10.4	12.63
	26	8.3	7.5	16.1	21.6	14.8	11.5	13.30
	27	11.0	10.5	17.2	23.0	17.1	17.3	16.02
	28	17.3	17.2	17.3	19.9	18.7	14.4	17.47
	29	14.2	13.5	11.3	9.2	8.2	4.2	10.10
	30	1.5	1.9	9.9	13.8	8.6	4.6	6.72
	31	6.3	7.8	11.7	15.8	12.7	10.3	10.77
	Mean	12.21	11.30	18.32	21.01	16.58	13.45	15.49

Month & Year	Day	2^h am	4^h am	6^h am	7^h am	8^h am	9^h am	10^h am	11^h am	Noon	1^h pm	2^h pm	3^h pm	4^h pm	5^h pm	6^h pm	8^h pm	10^h pm	Midnight	Mean
FUJI 3718m.																				
		mm	mm	mm	mm	mm	mm	mm	mm	mm	mm	mm	mm	mm	mm	mm	mm	mm	mm	mm
VIII 1880	4	..	..	6,3	5,3	6,4	6,5	8,0	6,7	6,4	6,0	6,0	6,4	5,9	6,0	5,6	..	..	..	..
	5	..	..	4,3	4,5	5,7	5,7	5,8	5,9	6,0	6,1	6,3	6,7	6,5	6,3	6,1	..	..	..	..
IX 1887	4	..	..	..	..	..	..	..	..	..	..	..	..	5,1	3,7	3,4	4,0	3,7	4,4	..
	5	1,8	3,3	3,4	3,7	4,0	3,9	3,9	3,3	3,5	3,3	3,7	..	2,5	..	3,7	..	3,9	..	..
VIII 1889	1	4,1	4,0	4,1	..	4,6	..	6,2	..	5,1	..	4,6	..	5,0	..	5,7	4,6	4,4	4,6	4,75
	2	4,8	4,9	5,1	..	5,0	..	7,0	..	5,6	..	6,3	..	5,8	..	5,1	5,1	5,6	5,6	5,49
	3	5,6	6,4	3,6	..	4,1	..	4,0	..	4,0	..	5,8	..	6,0	..	4,6	4,1	4,5	5,0	4,72
	4	3,0	2,8	4,6	..	4,2	..	4,6	..	6,0	..	6,3	..	1,5	..	1,6	4,5	4,6	4,5	4,02
	5	2,0	2,6	1,6	..	3,8	..	5,1	..	3,2	..	6,2	..	6,3	..	6,2	4,9	5,4	5,0	4,36
	6	4,6	5,9	5,0	..	6,3	..	6,0	..	4,0	..	5,1	..	5,2	..	4,1	2,4	2,5	2,2	4,44
	7	1,9	1,9	2,0	..	1,9	..	1,5	..	1,6	..	4,9	..	5,9	..	3,4	2,1	2,1	2,3	2,62
	8	2,5	4,4	2,4	..	5,6	..	7,1	7,8	5,5	6,5	5,7	6,1	6,5	7,3	5,2	7,1	5,8	6,9	5,47
	9	6,8	6,3	6,5	6,3	7,3	5,9	6,0	7,3	8,3	9,0	8,9	6,1	7,1	6,8	6,5	6,7	5,9	6,2	6,87
	10	5,7	4,1	4,8	4,2	4,2	4,7	5,3	4,2	4,6	5,5	7,5	6,8	7,0	6,2	5,5	4,4	6,5	6,5	5,51
	11	6,8	6,4	5,6	5,6	7,0	5,3	4,4	4,9	2,6	3,4	4,4	5,0	6,2	6,1	5,9	5,8	5,4	4,9	5,45
	12	4,2	2,6	2,9	4,8	5,6	4,7	4,2	3,6	5,0	4,4	5,6	5,8	6,1	5,8	5,4	5,7	5,7	6,3	4,86
	13	5,9	5,9	6,0	6,3	6,4	6,3	5,9	5,5	6,2	5,7	5,7	6,8	6,3	5,8	5,5	5,5	5,4	5,4	5,84
	14	5,8	5,7	5,4	5,8	5,0	6,7	6,8	6,5	6,5	7,1	6,0	6,1	6,6	6,4	6,8	6,5	6,1	5,2	6,04
	15	4,2	3,1	3,6	4,2	3,5	4,0	3,8	4,1	3,7	4,5	4,9	4,3	3,1	2,9	1,2	1,6	2,5	3,8	3,25
	16	3,5	4,2	4,5	5,0	5,1	5,0	4,7	5,3	5,5	4,4	4,9	7,3	6,8	5,1	4,6	4,2	4,5	4,7	4,77
	17	4,8	3,7	4,4	4,3	5,1	5,7	5,0	4,5	6,0	5,5	4,8	5,3	6,8	6,1	5,3	6,8	6,5	6,6	5,40
	18	5,9	5,7	5,7	5,6	6,0	6,0	6,0	6,1	6,5	6,7	7,1	6,7	7,0	7,0	7,1	7,0	7,2	7,2	6,52
	19	7,4	7,0	6,7	6,9	7,5	7,4	7,2	7,0	6,9	6,7	6,6	6,5	6,4	6,3	6,2	6,4	6,6	6,7	6,80
	20	5,9	6,6	7,0	7,0	7,2	7,2	7,0	7,1	7,2	7,2	7,3	7,3	7,5	7,7	7,2	7,0	7,3	7,4	7,05
	21	7,5	7,5	7,4	7,6	7,4	7,5	7,4	7,4	7,8	7,9	8,0	7,8	7,7	7,5	6,4	4,2	4,1	4,8	6,68
	22	4,6	5,4	5,8	6,2	6,4	6,4	7,1	5,9	6,2	6,8	7,0	7,8	8,2	6,6	6,1	5,6	5,2	4,9	6,04
	23	5,7	6,0	5,7	5,7	5,5	5,2	5,7	5,9	7,0	7,6	7,8	7,8	7,4	6,7	6,5	6,5	6,4	6,3	6,38
	24	6,8	5,4	5,4	5,5	6,6	7,4	6,6	9,1	8,8	8,5	7,8	8,3	9,1	7,1	6,1	5,5	5,0	4,5	6,38
	25	5,5	5,4	5,0	5,5	5,6	6,2	6,1	6,1	6,3	6,2	6,2	6,9	7,1	6,3	6,1	5,5	6,3	5,0	5,84
	26	3,2	2,9	5,2	3,9	4,9	3,9	3,4	3,4	3,8	5,8	5,2	7,3	7,2	7,8	7,5	7,2	6,3	6,7	5,29
	27	6,3	6,6	7,3	7,4	8,2	8,9	9,3	9,8	9,3	9,3	9,5	9,6	9,2	9,6	8,8	7,7	5,8	7,6	7,97
	28	7,3	6,9	7,4	7,7	8,4	8,8	8,4	8,6	8,7	8,5	8,1	8,4	8,0	8,0	7,7	7,9	7,2	7,1	7,76
	29	7,3	7,2	7,2	7,3	7,4	8,0	8,5	8,9	8,7	9,0	8,8	8,1	8,4	7,6	7,4	6,8	6,5	6,5	7,56
	30	6,3	6,0	5,8	6,0	5,4	5,8	3,2	3,2	3,0	2,9	2,3	1,9	1,9	2,2	3,5	2,6	1,2	1,1	3,52
	31	1,1	1,0	1,2	2,8	2,9	2,5	1,9	3,8	0,8	3,0	2,0	2,0	1,9	1,7	2,5	2,4	3,7	3,4	1,99
	Mean	5,03	4,95	5,03	5,72	5,59	5,11	5,66	6,08	5,63	6,34	6,17	6,50	6,30	6,28	5,54	5,30	5,28	5,25	5,47
IX 1891	1	4,1	..	5,1	..	..	..	2,6	..	..	..	6,1	..	..	..	5,8	..	6,0	..	4,95
	2	6,3	..	6,5	..	..	..	7,3	..	..	..	6,9	..	..	..	6,6	..	6,0	..	6,60
	3	6,4	..	6,9	..	..	..	5,9	..	..	..	5,6	..	..	..	3,0	..	2,7	..	5,08
	4	4,9	..	6,0	..	..	..	2,3	..	..	..	1,8	..	..	..	2,6	..	2,1	..	3,28
	5	6,7	..	6,8	..	..	..	4,4	..	..	..	7,0	..	..	..	5,4	..	6,1	..	5,90
	6	7,1	..	7,3	..	..	..	6,9	..	..	..	7,2	..	..	..	4,5	..	2,1	..	5,85
	7	1,4	..	2,9	..	..	..	1,7	..	..	..	1,8	..	..	..	5,8	..	6,8	..	3,23
General Mean		5,05	4,95	5,26	5,72	5,59	5,11	5,44	6,08	5,63	6,34	5,99	6,50	6,30	6,28	5,40	5,30	5,16	5,25	5,51
ONTAKE 3062m.																				
VIII 1891	1	3,0	2,8	3,1	..	2,9	..	5,4	..	5,1	..	6,4	..	6,0	..	6,1	6,1	8,0	8,0	5,24
	2	8,1	8,3	8,7	..	8,7	..	9,1	..	8,6	..	8,7	..	7,5	..	7,6	7,2	8,0	8,2	8,22
	3	8,1	8,5	8,4	..	8,2	..	8,4	..	8,7	..	8,5	..	8,1	..	7,1	7,5	7,6	7,5	8,05
	4	8,0	8,0	8,5	..	8,7	..	9,1	..	8,6	..	8,5	..	8,7	..	7,5	5,7	6,5	8,2	8,00
	5	8,0	8,1	8,1	..	9,0	..	8,7	..	9,2	..	9,9	..	8,5	..	8,5	8,1	8,0	8,0	8,51
	6	8,3	8,2	8,0	..	8,0	..	7,5	..	9,6	..	9,8	..	10,0	..	8,6	8,3	7,7	7,9	8,49
	7	8,1	8,2	7,9	..	9,3	..	9,5	..	9,3	..	9,9	..	8,6	..	7,9	8,1	4,8	3,6	7,93
	8	8,4	6,1	6,1	..	7,1	..	6,2	..	7,2	..	7,8	..	6,3	..	6,9	5,7	5,1	5,3	6,10
	9	4,9	5,3	5,6	..	6,2	..	8,7	..	8,4	..	9,1	..	8,2	..	6,8	5,2	8,0	8,2	7,05
	10	8,0	8,1	8,1	..	8,4	..	8,4	..	9,0	..	8,9	..	9,0	..	8,2	8,3	7,9	8,0	8,36
	11	7,5	6,4	7,1	..	8,2	..	8,4	..	8,7	..	8,9	..	9,0	..	8,5	7,4	8,0	7,6	7,97
	12	6,9	5,9	5,4	..	6,4	..	8,3	..	7,9	..	8,2	..	7,1	..	7,9	7,3	8,0	8,5	7,32
	13	8,5	8,0	8,1	..	8,1	..	8,5	..	10,0	..	9,5	..	9,2	..	7,6	6,9	6,8	6,4	8,13
	14	6,1	6,6	8,2	..	7,4	..	8,7	..	9,3	..	10,0	..	9,2	..	9,1	8,3	8,7	5,5	8,09
	15	5,0	5,4	6,5	..	4,4	..	5,2	..	7,5	..	8,1	..	8,8	..	6,2	6,4	8,1	6,4	6,50

Month & Year	Day	2h am	4h am	6h am	7h am	8h am	9h am	10h am	11h am	Noon	1h pm	2h pm	3h pm	4h pm	5h pm	6h pm	7h pm	8h pm	9h pm	10h pm	11h pm	Mid-night	Mean
HARA 3m.																							
		mm	mm	mm	mm	mm	mm	mm	mm	mm	mm	mm	mm	mm	mm	mm	mm	mm	mm	mm	mm	mm	mm
VIII 1880	4	..	..	18.0	18.8	19.8	20.7	19.4	20.0	20.7	20.7	22.5	23.6	23.8	24.1	22.5	..	..	..	..	..	..	..
	5	..	..	19.9	19.9	20.5	23.4	22.4	22.0	23.9	25.6	22.9	22.8	24.0	25.2	25.6							
TOKIO 21m.																							
IX 1887	4	..	..	..	..	..	..	..	..	..	..	..	..	17.0	17.1	17.4	17.4	17.3	16.6	16.5	16.7	16.8	..
	5	16.3	13.8	14.1	14.3	14.1	13.1	13.7	12.6	13.4	13.0	12.8	..	11.8		12.9	..			14.7			
YAMANAKA 990m.																							
VIII 1889	1	14.7	..	14.7	18.0	17.5	17.3	18.8	17.9	19.0	19.4	17.4	19.7	19.3	19.9	19.0	19.1	18.1	17.2	16.5	..	..	16.85
	2	15.7	..	15.0	17.7	18.7	18.5	18.3	19.2	18.3	19.2	18.9	18.5	17.0	16.9	16.0	14.7	15.6	15.9	16.0	..	..	16.65
	3	15.2	..	15.5	17.6	18.5	19.1	18.8	19.1	19.2	20.2	19.7	19.7	21.1	19.5	19.0	18.8	18.6	18.3	17.9	..	..	17.08
	4	16.0	..	15.8	17.8	18.2	17.6	17.6	17.9	19.0	16.9	17.3	16.6	17.2	16.0	16.0	16.7	16.6	16.3	15.5	..	..	16.37
	5	14.2	..	14.6	17.0	17.4	16.5	17.1	16.1	15.7	17.1	18.3	16.6	15.9	15.8	15.9	15.8	16.0	15.6	14.6	..	..	15.78
	6	13.9	..	14.4	16.0	17.7	16.6	18.7	18.6	17.8	18.1	17.8	18.2	18.4	18.4	18.4	18.4	18.0	18.0	17.2	..	..	16.73
	7	14.3	..	16.1	15.8	17.7	17.2	18.4	17.2	19.0	17.0	18.3	18.2	17.9	18.2	17.5	17.4	16.8	17.9	16.3	..	..	16.82
	8	16.2	..	15.2	15.8	15.7	16.8	18.5	17.4	17.1	18.2	17.5	17.6	17.8	17.2	17.7	16.9	16.4	16.8	16.6	..	..	16.95
	9	16.2	..	16.8	17.1	17.3	17.6	18.9	19.2	19.1	19.5	19.0	17.5	17.3	17.2	17.0	16.9	16.5	16.9	17.3	15.9	16.5	17.45
	10	15.9	..	15.2	17.5	16.7	18.1	17.6	17.9	18.4	18.2	18.8	19.1	18.8	18.4	18.6	17.6	17.7	18.1	17.7	17.2	17.8	17.30
	11	16.6	..	17.4	18.0	18.1	18.4	18.6	18.4	18.0	18.5	19.1	19.1	18.9	18.7	18.2	18.3	18.5	18.6	18.3	17.3	16.6	18.06
	12	16.6	..	17.3	17.9	17.7	17.8	18.0	17.3	20.7	18.1	18.2	17.8	17.4	17.7	17.4	16.9	16.4	16.9	15.9	16.3	16.1	17.23
	13	15.2	..	16.1	16.5	16.8	15.8	16.1	16.4	15.7	17.1	15.6	16.4	16.6	15.4	15.9	16.8	15.8	15.1	14.0	14.2	13.5	15.48
	14	12.5	..	13.7	15.2	15.9	15.9	16.0	17.2	16.7	17.1	17.0	15.6	13.4	15.9	15.4	15.6	15.9	15.4	15.3	14.6	13.7	14.98
	15	12.6	..	12.8	15.4	14.7	15.0	14.2	14.9	14.2	13.7	13.7	13.5	14.4	13.8	14.0	12.6	11.8	12.4	12.1	10.8	10.4	13.23
	16	9.8	..	12.0	13.8	14.5	14.1	13.9	14.1	15.4	13.9	13.7	12.8	11.9	13.2	14.5	15.0	14.0	14.0	14.3	..	13.6	13.03
	17	13.0	..	12.1	14.5	13.4	15.7	15.0	15.3	16.7	17.1	17.1	15.8	16.6	17.0	16.8	16.5	16.3	16.0	16.0	..	..	15.00
	18	15.7	..	15.1	14.4	14.4	15.0	15.2	15.7	15.1	15.4	15.8	16.3	15.9	17.7	17.3	18.0	17.1	17.5	18.1	18.1	18.1	16.20
	19	17.4	..	17.2	17.5	18.5	18.1	18.0	18.0	17.7	18.4	15.8	17.1	16.8	16.9	16.5	15.8	16.2	16.4	16.3	..	16.3	16.87
	20	16.0	..	16.3	16.6	16.6	17.0	17.5	17.7	17.9	17.5	17.4	16.9	17.4	17.2	16.8	17.2	17.2	17.2	17.4	17.2	17.0	16.90
	21	16.9	..	17.3	17.6	18.1	18.8	19.1	19.9	20.5	20.2	20.9	19.5	18.9	18.5	16.8	17.8	17.8	17.8	16.6	..	15.4	17.93
	22	16.0	..	15.7	17.8	17.8	18.2	18.9	18.2	19.5	19.2	19.5	18.7	18.2	18.7	18.8	17.9	17.7	17.8	16.0	..	16.3	17.48
	23	16.8	..	14.5	16.8	17.4	18.5	17.7	18.4	19.4	19.2	20.3	19.3	19.6	18.3	17.4	16.8	17.0	16.7	16.7	16.7	15.7	17.22
	24	14.4	..	13.7	15.3	17.0	17.9	17.4	18.5	19.7	20.6	18.1	18.2	18.0	17.9	17.9	17.8	16.5	16.5	15.9	16.6	16.2	16.23
	25	15.7	..	15.6	17.0	15.8	16.6	16.5	17.7	18.7	18.3	17.6	18.3	16.9	17.6	17.2	17.2	17.0	16.6	16.2	16.3	16.2	16.47
	26	14.9	..	14.9	15.6	15.5	16.5	16.8	15.4	14.8	13.8	14.3	13.1	15.2	16.0	15.9	15.9	15.5	15.8	16.7	16.5	16.7	15.58
	27	19.5	..	16.0	16.5	17.1	17.3	18.3	17.6	18.5	18.5	17.8	17.7	18.9	18.4	18.3	18.2	17.8	18.0	18.0	18.0	17.9	17.98
	28	18.1	..	16.3	16.5	16.3	16.7	16.5	16.2	16.8	16.4	16.0	16.3	15.2	15.4	14.8	14.3	14.2	14.0	14.3	13.9	13.6	16.00
	29	12.7	..	12.3	11.9	12.0	12.5	12.8	13.0	13.0	12.8	13.2	13.1	13.1	12.8	12.3	11.7	11.9	12.0	11.7	12.3	11.5	12.50
	30	11.3	..	11.5	11.9	12.0	12.5	12.4	13.2	13.3	12.8	12.5	12.6	12.7	12.3	12.5	12.1	12.4	12.5	12.4	12.0	11.2	12.10
	31	12.2	..	10.9	11.4	11.6	11.8	12.7	13.3	13.7	14.3	13.2	12.6	12.8	12.7	12.4	12.4	12.3	11.7	11.3	11.3	11.5	12.12
	Mean	15.01	..	14.89	16.08	16.34	16.63	16.91	17.00	17.37	17.34	17.09	16.85	16.95	16.79	16.52	16.36	16.09	16.13	15.78	15.29	15.08	16.04
IX 1889	1	10.8	..	10.7	..	..	..	12.6	..	..	..	11.0	..	..	..	11.6	..	..	..	14.1	..	..	11.30
	2	12.4	..	11.1	..	..	..	15.5	..	..	..	17.4	..	..	..	16.1	..	..	..	16.7	..	..	14.70
	3	15.7	..	15.1	..	..	..	16.1	..	..	..	16.9	..	..	..	18.8	..	..	..	13.3	..	..	15.98
	4	12.5	..	11.9	..	..	..	13.6	..	..	..	12.3	..	..	..	12.4	..	..	..	12.2	..	..	12.50
	5	12.2	..	12.1	..	..	..	13.8	..	..	..	15.4	..	..	..	14.4	..	..	..	14.0	..	..	13.65
	6	14.2	..	15.2	..	..	..	17.8	..	..	..	18.0	..	..	..	18.1	..	..	..	15.8	..	..	16.52
	7	14.9	..	15.0	..	..	..	14.9	..	..	..	13.9	..	..	..	13.5	..	..	..	13.0	..	..	14.20
General Mean		16.03	..	14.54	16.08	16.34	16.63	16.54	17.00	17.37	17.34	16.70	16.85	16.95	16.79	16.24	16.36	16.09	16.13	15.46	15.29	15.08	15.60
KUROSAWA 832m.																							
VIII 1891	1	11.2	9.9	9.7	..	11.8	..	12.1	..	11.8	..	12.2	..	13.3	..	14.6	..	14.8	..	14.7	..	14.5	12.55
	2	14.5	14.0	14.4	..	15.1	..	16.4	..	17.5	..	17.0	..	14.0	..	15.3	..	15.0	..	14.8	..	14.4	15.20
	3	14.1	14.3	14.6	..	15.6	..	16.9	..	16.0	..	14.7	..	15.1	..	14.5	..	14.7	..	14.2	..	14.3	14.92
	4	14.3	14.8	16.4	..	17.5	..	16.9	..	18.8	..	16.5	..	17.8	..	17.3	..	16.2	..	15.3	..	14.7	16.35
	5	14.8	14.8	15.1	..	15.3	..	16.5	..	16.7	..	17.6	..	17.7	..	17.4	..	17.3	..	16.7	..	16.3	16.35
	6	16.4	16.2	15.9	..	16.3	..	16.4	..	18.7	..	18.2	..	18.2	..	16.2	..	17.3	..	17.0	..	16.2	16.92
	7	16.1	15.9	16.0	..	16.8	..	16.9	..	16.5	..	18.6	..	19.8	..	18.8	..	17.9	..	15.8	..	15.0	17.01
	8	14.5	13.4	13.4	..	14.5	..	16.7	..	14.3	..	16.4	..	13.7	..	15.1	..	14.9	..	14.9	..	14.3	14.67
	9	14.2	13.6	13.5	..	14.8	..	15.9	..	17.6	..	17.1	..	14.6	..	17.2	..	16.6	..	15.6	..	15.2	15.49
	10	14.8	14.6	14.6	..	16.7	..	16.8	..	16.6	..	15.7	..	16.4	..	17.5	..	16.7	..	16.1	..	15.9	16.03
	11	15.8	14.7	14.8	..	16.8	..	16.8	..	17.3	..	15.4	..	16.5	..	16.0	..	16.8	..	16.3	..	15.2	16.08
	12	14.8	14.0	14.0	..	15.5	..	13.4	..	17.6	..	16.7	..	14.6	..	16.7	..	17.0	..	15.2	..	15.5	15.42
	13	15.4	14.5	14.1	..	16.2	..	17.6	..	16.9	..	19.9	..	18.8	..	15.8	..	15.7	..	15.3	..	13.8	16.17
	14	13.4	11.5	12.9	..	15.5	..	16.3	..	16.2	..	16.8	..	18.2	..	18.3	..	16.5	..	17.9	..	17.0	15.87
	15	16.9	16.1	15.5	..	16.4	..	16.5	..	15.0	..	15.8	..	15.2	..	15.4	..	16.7	..	16.1	..	14.5	15.84

TABLE III. ABSOLUTE HUMIDITY.

Month & Year	Day	2h am	4h am	6h am	7h am	8h am	9h am	10h am	11h am	Noon	1h pm	2h pm	3h pm	4h pm	5h pm	6h pm	8h pm	10h pm	Midnight	Mean
ONTAKE 3062m.																				
		mm	mm	mm	mm	mm	mm	mm	mm	mm	mm	mm	mm	mm	mm	mm	mm	mm	mm	mm
VIII 1891	16	7.1	7.8	6.9	..	7.7	..	8.2	..	8.2	..	8.5	..	8.2	..	8.3	7.8	8.2	8.8	7.97
	17	8.7	8.7	8.7	..	9.9	..	8.4	..	9.0	..	8.7	..	8.8	..	8.5	8.8	9.0	8.6	8.74
	18	8.5	8.7	8.5	..	8.6	..	8.7	..	9.0	..	8.7	..	8.8	..	8.3	5.3	6.3	4.7	7.84
	19	3.8	3.0	4.1	..	3.5	..	6.6	..	4.8	..	5.3	..	6.4	..	3.3	1.5	2.6	1.9	3.90
	20	1.8	1.3	1.1	..	2.9	..	3.2	..	3.4	..	5.9	..	5.1	..	2.6	1.4	0.9	1.7	2.61
	21	1.5	6.4	6.7	..	6.4	..	7.6	..	8.0	..	8.0	..	7.5	..	7.9	7.8	7.9	8.2	6.99
	22	8.3	7.4	6.7	..	7.2	..	8.1	..	9.1	..	8.3	..	7.6	..	7.6	7.7	7.8	7.7	7.79
	23	7.5	7.1	7.6	..	7.9	..	8.5	..	8.4	..	8.0	..	7.4	..	6.6	5.5	6.4	5.2	7.17
	24	5.3	5.4	5.7	..	6.0	..	6.8	..	8.1	..	7.6	..	7.0	..	5.7	5.9	5.9	5.6	6.25
	25	5.1	5.4	5.8	..	5.8	..	5.2	..	6.9	..	8.0	..	6.5	..	3.9	4.0	3.9	2.0	5.21
	26	1.5	2.3	1.9	..	3.7	..	4.9	..	7.4	..	7.2	..	6.9	..	3.4	2.8	1.1	1.0	3.67
	27	1.1	0.8	1.1	..	3.8	..	7.2	..	6.3	..	6.9	..	5.8	..	6.2	5.9	4.4	2.2	4.31
	28	3.4	3.1	5.3	..	6.2	..	6.2	..	6.7	..	7.1	..	6.8	..	5.8	3.8	3.3	6.2	5.32
	29	6.1	6.5	6.4	..	6.7	..	7.6	..	4.8	..	6.7	..	6.6	..	4.8	2.3	0.1	0.3	4.91
	30	0.1	3.9	1.9	..	1.9	..	5.5	..	7.1	..	6.9	..	7.3	..	4.3	5.4	0.5	0.5	3.77
	31	2.1	2.0	2.5	..	4.3	..	5.9	..	8.0	..	8.2	..	8.1	..	4.0	3.6	4.0	3.1	4.65
	Mean	5.61	5.93	6.09	..	6.53	..	7.38	..	7.82	..	8.12	..	7.71	..	6.03	6.00	5.52	5.64	6.62
IX 1891	1	5.6	5.6	6.2	..	6.4	..	8.4	..	8.7	..	8.3	..	8.4	..	8.2	7.5	8.0	8.2	7.46
	2	8.2	8.1	7.4	..	8.5	..	8.9	..	9.1	..	9.3	..	8.8	..	8.6	8.1	7.8	7.5	8.36
	3	7.7	7.2	6.7	..	7.4	..	8.9	..	9.6	..	9.3	..	9.4	..	8.8	7.2	6.1	6.3	7.88
	4	4.8	4.3	6.2	..	6.6	..	8.1	..	9.8	..	8.7	..	8.7	..	7.2	5.2	4.7	4.7	6.58
	5	5.3	3.9	4.3	..	6.2	..	4.9	..	6.2	..	7.1	..	7.7	..	6.1	5.9	4.2	4.1	5.49
	6	3.8	6.5	7.0	..	7.9	..	7.8	..	9.2	..	8.6	..	8.7	..	5.5	5.4	5.9	3.0	6.61
	7	2.5	2.3	3.1	..	4.0	..	3.7	..	6.3	..	6.2	..	4.3	..	3.8	3.6	4.0	6.3	7.17
	8	7.2	7.3	6.6	..	6.6	..	7.7	..	7.6	..	7.8	..	6.8	..	7.3	7.0	7.4	8.0	7.27
	9	7.9	8.1	8.4	..	8.7	..	8.9	..	9.1	..	9.0	..	8.8	..	8.1	8.4	8.5	7.9	8.48
	10	7.8	7.8	7.2	..	5.4	..	4.8	..	7.7	..	8.1	..	8.7	..	9.1	9.2	9.3	7.9	7.75
	11	8.0	7.8	8.8	..	8.5	..	9.1	..	9.0	..	9.9	..	8.8	..	8.1	7.2	8.1	7.5	8.32
	12	7.4	7.1	6.9	..	7.9	..	9.5	..	10.2	..	7.7	..	8.1	..	8.3	8.3	8.7	8.5	8.22
General Mean		5.81	6.04	6.22	..	6.66	..	7.48	..	8.02	..	8.17	..	7.82	..	6.86	6.26	6.12	5.93	6.78
GOZAISHODAKE 1200m.																				
IX 1888	4	16.3	15.8	16.1	..	15.8	..	16.9	..	15.4	..	15.8	..	15.8	..	15.6	14.4	14.3	14.3	15.4
	5	13.9	13.0	12.2	..	12.6	..	12.5	..	12.1	..	11.5	..	11.5	..	9.9	9.6	9.2	9.0	11.5
	6	9.0	8.7	8.2	..	8.2	..	8.2	..	9.2	..	10.6	..	11.8	..	9.5	8.8	10.5	10.7	9.4
	7	10.9	10.3	10.9	..	11.8	..	11.5	..	12.2	..	11.7	..	11.1	..	11.3	8.1	9.4	10.9	10.8
	8	11.2	12.5	12.8	..	13.2	..	13.3	..	13.3	..	12.8	..	11.0	..	9.7	11.0	10.8	9.9	11.8
	9	10.5	11.3	11.6	..	12.3	..	13.9	..	12.7	..	13.8	..	14.1	..	13.4	12.5	12.4	12.9	12.5
	10	13.2	12.4	12.8	..	13.5	..	13.9	..	14.4	..	15.1	..	14.8	..	14.5	13.9	12.4	12.1	13.6
	11	11.9	11.9	11.9	..	11.9	..	12.4	..	12.5	..	12.7	..	12.9	..	13.0	13.1	13.1	13.7	12.5
	12	15.4	14.9	13.7	..	14.9	..	14.7	..	15.0	..	14.7	..	13.2	..	12.4	11.9	11.6	10.9	13.6
	13	9.8	8.9	8.3	..	5.6	..	10.7	..	10.6	..	9.2	..	8.4	..	8.5	8.4	8.4	8.4	8.8
	14	8.8	8.7	8.5	..	9.2	..	9.2	..	9.8	..	10.7	..	9.1	..	8.3	8.2	7.7	7.1	8.8
	15	7.0	7.1	7.4	..	8.8	..	8.4	..	11.7	..	12.1	..	12.2	..	11.4	11.2	11.8	10.7	10.0
	16	10.3	10.1	10.6	..	10.9	..	11.9	..	12.9	..	13.9	..	12.2	..	11.4	11.4	10.2	11.5	11.6
	17	10.8	10.6	10.2	..	10.6	..	10.6	..	11.0	..	9.9	..	9.5	..	9.6	8.5	7.1	7.7	9.7
	18	9.0	9.9	10.2	..	10.5	..	9.6	..	8.8	..	10.1	..	9.0	..	9.5	8.3	7.2	6.8	9.0
	19	5.8	6.8	10.1	..	10.8	..	11.4	..	11.7	..	11.6	..	11.6	..	11.6	11.6	12.1	12.1	10.5
	20	12.1	12.5	11.9	..	12.2	..	12.6	..	13.0	..	13.0	..	13.2	..	12.9	13.3	13.2	12.9	12.8
	21	12.6	12.4	12.1	..	12.6	..	12.9	..	13.0	..	12.6	..	12.4	..	11.6	12.1	11.9	11.6	12.3
	22	12.0	12.2	11.4	..	11.6	..	12.4	..	13.3	..	12.5	..	12.6	..	12.2	12.2	11.7	10.9	12.0
	23	10.7	10.7	10.7	..	11.3	..	11.7	..	11.7	..	11.2	..	12.1	..	10.6	10.8	11.1	8.1	10.9
	24	5.4	3.4	3.7	..	5.4	..	7.9	..	11.2	..	11.1	..	10.5	..	9.2	8.3	7.9	7.5	7.6
	25	5.2	2.0	1.6	..	4.8	..	10.6	..	10.4	..	10.9	..	11.0	..	10.7	10.9	11.1	10.7	8.3
	26	10.3	10.3	10.5	..	11.4	..	12.3	..	12.8	..	12.9	..	12.0	..	10.5	10.9	10.3	10.2	11.2
	27	10.7	10.3	10.4	..	10.6	..	11.0	..	12.1	..	11.8	..	8.3	..	9.9	10.4	10.3	9.4	10.4
	28	2.6	3.2	5.8	..	5.6	..	4.6	..	7.6	..	9.7	..	9.4	..	8.3	7.8	8.1	8.2	6.7
	29	9.2	8.9	8.4	..	9.7	..	10.6	..	9.7	..	10.4	..	9.9	..	9.6	8.5	7.3	6.6	9.1
	30	6.9	6.7	6.4	..	8.4	..	8.4	..	9.1	..	8.3	..	7.7	..	9.0	6.9	6.7	7.3	7.7
X 1888	1	6.8	7.2	5.6	..	7.3	..	8.5	..	8.3	..	8.0	..	6.6	..	5.6	5.2	5.4	6.7	6.7
	2	6.8	6.5	5.7	..	4.8	..	5.8	..	5.9	..	6.2	..	6.3	..	5.9	6.4	5.8	5.7	6.0
	3	5.4	5.3	5.2	..	5.4	..	5.2	..	5.5	..	5.4	..	6.5	..	5.9	5.7	4.4	..	5.0
	Mean	9.7	9.5	9.5	..	10.1	..	10.7	..	11.2	..	11.3	..	10.9	..	10.4	10.0	9.8	9.5	10.2

Month & Year	Day	2h am	4h am	6h am	7h am	8h am	9h am	10h am	11h am	Noon	1h pm	2h pm	3h pm	4h pm	5h pm	6h pm	7h pm	8h pm	9h pm	10h pm	11h pm	Mid-night	Mean
		KUROSAWA 832m.																					
		mm	mm	mm	mm	mm	mm	mm	mm	mm	mm	mm	mm	mm	mm	mm	mm	mm	mm	mm	mm	mm	mm
VIII 1891	16	13.7	12.4	12.7	..	14.1	..	12.7	..	13.2	..	12.6	..	12.7	..	14.4	..	15.7	..	15.7	..	15.5	13.81
	17	15.5	15.7	15.0	..	16.1	..	16.6	..	15.1	..	15.4	..	17.6	..	17.5	..	17.4	..	17.1	..	16.5	16.29
	18	16.5	16.6	16.6	..	17.1	..	18.8	..	16.5	..	17.2	..	17.5	..	15.9	..	14.0	..	13.7	..	12.5	16.07
	19	12.2	11.3	10.7	..	12.5	..	14.0	..	12.6	..	12.2	..	11.6	..	10.0	..	10.6	..	11.7	..	10.6	11.67
	20	10.0	9.3	9.6	..	10.4	..	10.9	..	9.9	..	12.0	..	13.5	..	17.4	..	13.2	..	11.7	..	11.6	11.62
	21	11.7	11.7	11.1	..	11.6	..	12.4	..	13.3	..	15.2	..	15.6	..	15.5	..	16.3	..	15.5	..	15.8	13.72
	22	15.6	14.8	14.8	..	15.7	..	16.5	..	16.6	..	16.2	..	16.6	..	14.3	..	15.2	..	15.0	..	14.8	15.42
	23	14.6	14.5	14.8	..	14.9	..	15.5	..	16.0	..	15.3	..	15.6	..	15.3	..	14.6	..	14.0	..	13.0	14.82
	24	12.5	12.2	12.0	..	12.9	..	14.0	..	15.7	..	14.6	..	15.3	..	15.4	..	14.8	..	14.3	..	14.0	13.97
	25	13.2	12.8	12.7	..	13.7	..	13.8	..	12.1	..	12.2	..	16.5	..	14.9	..	13.4	..	12.2	..	12.1	13.30
	26	11.1	11.1	10.8	..	10.8	..	12.8	..	11.9	..	10.6	..	11.8	..	12.9	..	10.4	..	10.1	..	9.9	11.18
	27	8.9	8.6	8.5	..	10.8	..	11.8	..	14.1	..	11.8	..	14.0	..	14.7	..	13.5	..	12.5	..	11.6	11.56
	28	11.7	10.3	10.4	..	12.2	..	13.5	..	13.1	..	12.8	..	12.3	..	14.1	..	13.6	..	12.3	..	11.7	12.32
	29	11.6	11.3	11.1	..	12.0	..	13.4	..	13.6	..	12.4	..	11.4	..	13.4	..	12.4	..	11.9	..	10.8	12.11
	30	10.0	9.1	9.1	..	10.0	..	12.2	..	10.6	..	11.1	..	14.1	..	13.5	..	11.8	..	11.2	..	10.6	11.11
	31	9.6	9.5	9.6	..	11.5	..	13.6	..	13.4	..	14.8	..	15.6	..	16.3	..	14.7	..	13.9	..	13.4	12.97
	Mean	13.51	13.01	13.01	..	14.23	..	14.97	..	15.02	..	15.08	..	15.31	..	15.56	..	14.99	..	14.47	..	13.91	14.42
IX 1891	1	12.7	11.9	11.8	..	14.1	..	15.6	..	17.1	..	14.2	..	14.7	..	16.8	..	16.2	..	15.2	..	14.3	14.54
	2	13.7	13.5	13.5	..	15.5	..	17.0	..	16.5	..	19.1	..	16.4	..	17.6	..	16.3	..	15.4	..	14.8	15.77
	3	13.9	13.9	13.6	..	14.6	..	16.3	..	15.7	..	18.0	..	17.0	..	17.6	..	17.0	..	15.8	..	16.0	15.78
	4	15.4	15.1	14.3	..	15.3	..	17.4	..	17.4	..	16.2	..	16.1	..	15.8	..	15.5	..	14.8	..	14.0	15.61
	5	13.4	13.2	12.9	..	14.2	..	14.4	..	12.7	..	15.9	..	15.9	..	15.4	..	15.0	..	14.0	..	12.8	14.15
	6	12.7	12.1	11.7	..	13.0	..	14.9	..	14.7	..	15.6	..	14.3	..	13.9	..	13.7	..	13.5	..	12.1	13.52
	7	11.7	11.0	11.1	..	11.9	..	13.9	..	13.7	..	14.7	..	14.8	..	14.8	..	14.8	..	14.1	..	14.3	13.42
	8	14.1	14.0	12.1	..	14.2	..	15.5	..	15.9	..	15.5	..	17.7	..	16.5	..	15.5	..	15.0	..	15.1	15.09
	9	15.2	15.3	15.4	..	15.6	..	16.2	..	18.3	..	17.1	..	18.4	..	17.5	..	16.5	..	16.7	..	16.2	16.56
	10	16.1	15.5	15.4	..	17.1	..	18.5	..	14.1	..	15.0	..	15.6	..	16.7	..	17.6	..	17.4	..	15.8	15.82
	11	14.8	14.0	14.8	..	16.1	..	17.1	..	13.8	..	15.8	..	15.5	..	17.7	..	16.5	..	15.3	..	14.2	15.47
	12	14.0	13.3	12.7	..	15.2	..	17.5	..	15.0	..	15.0	..	15.1	..	16.9	..	15.7	..	16.2	..	16.1	15.22
General Mean		13.66	13.17	13.11	..	14.37	..	15.49	..	15.12	..	15.29	..	15.49	..	15.80	..	15.23	..	14.71	..	14.11	14.60
		YOKKAICHI 4m.																					
IX 1888	4	21.9	..	23.5	..	..	..	22.0	..	..	..	21.9	..	..	..	21.8	..	..	..	20.1	..	..	21.4
	5	19.9	..	21.8	..	..	..	18.5	..	..	..	15.3	..	..	..	13.6	..	..	..	12.5	..	..	16.9
	6	12.2	..	11.8	..	..	..	14.5	..	..	..	12.6	..	..	..	13.2	..	..	..	14.6	..	..	13.1
	7	15.7	..	15.5	..	..	..	16.6	..	..	..	16.6	..	..	..	17.8	..	..	..	16.8	..	..	16.5
	8	17.0	..	16.4	..	..	..	16.7	..	..	..	17.9	..	..	..	18.7	..	..	..	17.7	..	..	17.4
	9	17.5	..	16.2	..	..	..	16.9	..	..	..	16.8	..	..	..	19.3	..	..	..	17.4	..	..	17.3
	10	17.4	..	16.7	..	..	..	18.6	..	..	..	19.1	..	..	..	18.8	..	..	..	16.0	..	..	17.8
	11	15.9	..	16.7	..	..	..	17.5	..	..	..	18.6	..	..	..	17.6	..	..	..	20.0	..	..	17.7
	12	19.3	..	21.2	..	..	..	21.3	..	..	..	22.3	..	..	..	18.1	..	..	..	17.0	..	..	19.9
	13	13.8	..	13.1	..	..	..	17.1	..	..	..	12.9	..	..	..	8.7	..	..	..	11.9	..	..	12.9
	14	11.6	..	11.8	..	..	..	13.9	..	..	..	12.6	..	..	..	12.3	..	..	..	13.2	..	..	12.7
	15	11.4	..	12.0	..	..	..	12.9	..	..	..	13.2	..	..	..	15.1	..	..	..	16.6	..	..	13.5
	16	14.9	..	15.3	..	..	..	16.4	..	..	..	16.0	..	..	..	16.6	..	..	..	15.7	..	..	15.8
	17	15.0	..	13.9	..	..	..	14.2	..	..	..	14.6	..	..	..	13.4	..	..	..	13.0	..	..	14.0
	18	11.7	..	12.6	..	..	..	15.6	..	..	..	13.5	..	..	..	12.3	..	..	..	10.1	..	..	12.6
	19	10.7	..	11.5	..	..	..	13.6	..	..	..	15.0	..	..	..	17.0	..	..	..	17.9	..	..	14.3
	20	17.0	..	17.3	..	..	..	21.4	..	..	..	19.0	..	..	..	19.2	..	..	..	19.9	..	..	19.0
	21	19.0	..	17.1	..	..	..	17.4	..	..	..	18.7	..	..	..	18.1	..	..	..	18.8	..	..	18.2
	22	18.3	..	16.9	..	..	..	18.0	..	..	..	18.1	..	..	..	17.9	..	..	..	17.1	..	..	17.7
	23	14.4	..	14.1	..	..	..	16.3	..	..	..	16.6	..	..	..	15.0	..	..	..	13.7	..	..	15.1
	24	12.4	..	11.4	..	..	..	14.3	..	..	..	14.8	..	..	..	12.9	..	..	..	12.1	..	..	13.0
	25	10.7	..	9.8	..	..	..	11.3	..	..	..	14.0	..	..	..	12.3	..	..	..	15.1	..	..	12.2
	26	14.5	..	13.9	..	..	..	16.4	..	..	..	15.6	..	..	..	16.4	..	..	..	15.5	..	..	15.4
	27	15.9	..	14.4	..	..	..	16.3	..	..	..	16.2	..	..	..	16.7	..	..	..	15.1	..	..	15.8
	28	17.9	..	10.5	..	..	..	14.3	..	..	..	11.1	..	..	..	13.2	..	..	..	11.8	..	..	13.1
	29	12.6	..	13.2	..	..	..	14.1	..	..	..	14.6	..	..	..	13.3	..	..	..	11.9	..	..	13.0
	30	12.7	..	12.5	..	..	..	12.9	..	..	..	12.1	..	..	..	12.4	..	..	..	11.0	..	..	12.3
X 1888	1	11.1	..	11.5	..	..	..	11.5	..	..	..	12.2	..	..	..	9.4	..	..	..	8.0	..	..	10.6
	2	7.7	..	7.7	..	..	..	8.1	..	..	..	8.7	..	..	..	8.6	..	..	..	7.8	..	..	8.1
	3	6.4	..	7.6	..	..	..	7.7	..	..	..	7.7	..	..	..	8.6	..	..	..	7.7	..	..	7.6
	Mean	14.5	..	14.1	..	..	..	15.5	..	..	..	15.2	..	..	..	15.0	..	..	..	14.5	..	..	14.8

TABLE III. ABSOLUTE HUMIDITY.

Month & Year	Day	2^h am	6^h am	10^h am	2^h pm	6^h pm	10^h pm	Mean
		HIGASHI HOBEN 736m.						
		mm	mm	mm	mm	mm	mm	mm
VIII 1889	1	18.1	18.1	21.1	20.2	20.5	18.1	19.35
	2	18.8	18.2	21.2	22.9	20.9	20.0	20.42
	3	18.6	18.6	18.2	13.9	18.5	15.4	18.20
	4	14.2	13.9	15.6	16.1	15.8	13.4	14.83
	5	13.0	15.7	19.9	22.4	19.8	15.0	17.63
	6	18.3	19.0	19.4	20.2	19.6	20.1	19.48
	7	18.4	18.6	19.7	18.2	20.0	16.2	18.50
	8	18.2	17.7	19.0	19.2	17.6	11.1	17.13
	9	15.9	17.3	20.1	21.0	18.5	18.2	18.50
	10	16.9	16.5	20.3	20.2	20.1	14.3	18.05
	11	14.7	15.2	18.9	21.0	17.1	15.8	17.12
	12	15.8	15.8	17.8	17.0	16.6	14.5	16.25
	13	17.4	14.1	20.7	20.5	17.9	15.8	17.73
	14	16.7	18.4	17.3	15.2	15.7	16.7	16.50
	15	15.9	16.1	15.4	15.5	16.8	14.7	15.78
	16	16.2	15.5	16.5	17.1	15.7	17.0	16.33
	17	18.1	17.7	18.1	17.6	17.9	17.1	17.75
	18	17.3	16.6	18.7	18.4	16.8	17.3	17.52
	19	18.4	17.9	18.0	16.8	16.8	17.2	17.52
	20	16.9	16.9	17.9	18.3	17.9	17.8	17.62
	21	17.1	17.5	20.4	20.4	20.3	18.6	19.05
	22	16.2	18.5	20.5	21.8	20.2	18.5	19.28
	23	19.4	18.2	20.3	21.2	20.3	19.4	19.80
	24	19.2	19.1	19.2	18.7	18.8	18.0	18.83
	25	18.6	18.7	19.1	19.4	18.7	18.3	18.80
	26	17.2	16.0	15.9	15.7	14.8	14.7	15.72
	27	14.3	13.9	14.3	15.0	14.6	14.0	14.35
	28	14.1	11.9	12.9	14.1	13.2	13.5	13.28
	29	13.1	13.0	13.6	14.0	13.9	14.2	13.63
	30	12.1	12.9	15.1	14.7	14.7	15.0	14.08
	31	13.4	14.6	15.2	15.8	15.6	14.5	14.85
	Mean	16.55	16.53	18.06	18.34	17.59	16.24	17.22
IX 1889	1	14.2	14.7	15.8	18.2	17.1	17.2	16.20
	2	17.0	17.0	17.1	15.8	15.0	14.7	16.10
	3	13.6	11.9	14.9	15.0	14.3	14.8	14.08
	4	13.9	12.1	13.5	16.1	16.0	15.5	14.57
	5	15.9	16.6	16.9	18.9	18.3	17.5	17.35
	6	17.7	16.1	15.8	15.1	14.5	13.9	15.52
	7	13.7	14.3	14.3	14.7	14.7	13.8	14.25
	8	14.2	13.3	13.9	15.4	14.8	12.2	13.97
	9	11.1	10.2	9.9	14.3	13.2	13.2	11.98
	10	10.6	11.3	10.8	12.7	13.7	13.2	12.05
	11	13.3	13.3	14.3	13.0	12.6	12.5	13.17
	12	12.8	12.5	12.4	12.3	11.4	11.0	12.07
	13	10.9	10.2	11.7	12.1	12.0	8.5	10.90
	14	10.9	8.7	10.6	9.8	11.8	11.3	10.52
	15	11.5	10.1	10.3	12.2	11.3	10.8	11.03
	16	10.5	9.6	10.9	10.9	9.8	9.9	10.27
	17	9.2	9.0	10.6	10.7	10.6	10.2	10.05
	18	10.2	9.9	13.6	11.5	9.2	10.4	10.80
	19	9.6	12.2	12.1	12.4	12.9	13.0	12.03
	20	13.3	14.1	14.7	15.1	15.6	16.3	14.85
	21	16.0	16.3	15.9	15.6	13.3	12.7	14.97
	22	13.2	11.4	12.1	11.2	9.9	10.9	11.45
	23	11.0	10.0	11.1	12.1	11.7	10.5	11.07
	24	9.9	10.4	11.6	11.4	10.2	10.1	10.60
	25	10.3	9.4	9.9	10.3	12.1	11.4	10.57
	26	10.6	10.3	11.7	12.0	12.6	11.9	11.52
	27	12.6	12.2	11.7	13.9	12.8	12.2	12.57

TABLE III^c. ABSOLUTE HUMIDITY.

Month & Year	Day	2^h am	6^h am	10^h am	2^h pm	6^h pm	10^h pm	Mean
		YAMAGUCHI 35m.						
		mm	mm	mm	mm	mm	mm	mm
VIII 1889	1	20.3	20.2	23.5	21.4	21.6	21.2	21.37
	2	21.0	22.0	24.3	24.2	22.4	22.3	22.70
	3	19.7	19.3	22.2	22.1	21.4	20.3	20.83
	4	18.5	18.1	20.3	20.6	21.0	19.6	19.68
	5	18.8	17.8	20.6	21.4	22.6	20.8	20.33
	6	19.8	20.4	20.7	20.6	23.5	23.3	21.38
	7	21.5	21.7	19.9	19.4	20.9	20.1	20.58
	8	19.3	19.5	20.8	18.7	21.0	18.3	19.60
	9	17.5	17.1	20.9	19.4	22.0	20.8	19.62
	10	19.3	18.5	22.5	21.0	22.0	19.5	20.47
	11	20.0	18.6	21.3	19.1	20.7	18.7	19.73
	12	17.5	17.3	21.8	21.7	19.4	20.0	19.62
	13	17.8	17.8	21.6	19.3	20.5	19.3	19.38
	14	18.5	19.9	19.5	15.0	17.3	17.0	17.87
	15	17.6	18.8	16.0	17.8	18.2	15.9	17.37
	16	17.9	17.6	18.5	19.3	17.5	18.2	18.17
	17	20.6	20.7	20.0	19.2	19.1	18.2	19.63
	18	18.6	18.2	20.1	18.6	18.7	18.6	18.85
	19	19.3	19.3	20.5	19.1	18.5	18.2	19.12
	20	19.7	20.1	21.6	19.5	21.1	20.1	20.35
	21	18.4	18.7	21.0	21.4	21.4	20.5	20.23
	22	19.0	19.8	21.4	21.9	22.8	22.2	21.18
	23	22.3	22.6	22.2	21.6	22.1	22.1	22.15
	24	21.4	21.5	20.7	19.9	20.6	20.6	20.78
	25	21.0	21.2	21.9	22.9	22.4	22.2	21.93
	26	19.8	17.8	16.6	17.9	16.5	15.9	17.38
	27	15.8	15.6	16.7	20.9	16.0	16.3	16.88
	28	15.9	15.2	14.0	15.2	16.6	15.2	15.46
	29	14.1	14.4	15.0	15.3	15.3	15.5	14.98
	30	15.3	15.3	17.0	16.7	16.6	16.0	16.15
	31	16.4	16.5	17.3	18.6	20.0	18.0	17.80
	Mean	18.80	18.76	20.04	19.67	19.98	19.19	19.41
IX 1889	1	17.6	17.7	18.3	18.6	18.9	19.1	18.37
	2	19.2	19.7	20.0	17.7	17.3	16.4	18.38
	3	15.8	14.3	16.0	18.1	17.0	17.1	16.38
	4	16.3	15.6	15.9	16.1	17.5	18.0	16.57
	5	17.8	18.2	19.1	20.4	19.7	20.5	19.28
	6	21.1	21.1	17.9	17.7	15.8	15.8	18.23
	7	15.5	15.1	16.3	15.8	15.8	15.4	15.65
	8	13.9	13.5	16.2	18.7	17.0	14.6	15.65
	9	14.3	13.9	12.9	14.1	15.1	14.3	14.10
	10	14.3	13.1	13.7	15.3	15.6	16.3	14.72
	11	15.1	14.6	16.2	14.8	14.3	13.9	14.82
	12	14.4	14.2	14.6	13.5	13.6	13.0	13.88
	13	12.6	11.3	14.1	13.5	12.4	12.5	12.73
	14	11.1	10.8	11.0	9.6	12.2	12.7	11.23
	15	11.0	10.4	14.4	13.6	13.2	11.8	12.40
	16	11.9	9.2	12.0	11.0	12.1	11.7	11.17
	17	11.0	10.3	11.3	12.1	12.5	12.6	11.63
	18	11.1	9.9	12.8	12.6	12.4	12.6	11.90
	19	11.7	12.0	13.7	13.0	16.2	15.3	13.65
	20	15.9	16.3	17.0	18.2	18.0	18.9	17.38
	21	18.7	18.8	19.6	19.6	17.1	15.3	18.18
	22	14.4	14.5	14.8	14.0	13.6	12.3	13.93
	23	12.6	12.6	12.9	13.2	14.3	12.2	12.97
	24	11.0	11.1	13.5	14.1	12.6	11.0	12.22
	25	10.5	11.0	13.3	12.6	13.3	12.6	12.22
	26	10.7	10.3	13.2	13.8	14.9	14.6	12.92
	27	14.9	15.1	14.3	16.2	15.9	14.9	15.22

TABLE III. ABSOLUTE HUMIDITY

Month & Year	Day	2h am	6h am	10h am	2h pm	6h pm	10h pm	Mean
HIGASHI HOBEN 736m.								
IX 1889		mm	mm	mm	mm	mm	mm	mm
	28	11.8	11.9	12.8	11.3	11.7	11.2	11.78
	29	10.9	10.7	9.9	9.7	9.8	8.5	9.92
	30	7.9	6.6	6.4	6.8	7.2	7.2	7.02
	Mean	12.28	11.88	12.57	13.08	12.67	12.22	12.44
X 1889	1	7.0	6.5	6.6	7.2	6.6	6.1	6.67
	2	3.8	5.0	6.9	7.5	9.4	8.8	6.90
	3	7.9	8.0	8.6	10.1	9.8	8.0	8.73
	4	9.5	9.4	9.9	10.0	7.8	8.6	9.20
	5	1.7	5.9	8.6	8.1	11.0	10.9	7.76
	6	12.4	11.1	12.5	12.2	11.8	11.6	11.93
	7	11.4	10.7	10.8	10.4	10.3	9.0	10.43
	8	9.2	9.1	9.7	8.9	8.0	6.5	8.57
	9	8.3	7.1	6.7	7.8	7.3	4.9	7.02
	10	8.1	10.0	6.6	7.4	9.4	5.3	7.80
	11	4.5	6.9	7.7	6.3	6.9	7.6	6.65
	12	6.7	7.3	8.0	10.2	7.6	8.8	8.10
	13	8.0	7.4	7.3	4.4	7.0	5.0	6.62
	14	5.0	4.5	7.3	8.0	8.3	10.1	7.20
	15	10.6	11.3	12.5	12.5	12.2	12.5	11.93
	16	12.0	10.8	8.8	10.6	10.8	9.3	10.38
	17	9.4	9.2	8.8	9.7	9.4	6.6	8.85
	18	6.0	9.5	8.6	8.2	10.2	8.3	8.47
	19	8.5	6.2	8.1	8.1	4.1	6.6	6.93
	20	7.4	7.7	9.6	11.4	12.1	13.0	10.20
	21	14.9	14.1	11.9	8.3	6.8	5.4	10.28
	22	5.8	5.6	5.5	5.3	5.0	5.1	5.38
	23	4.4	5.0	5.6	5.4	5.7	5.8	5.30
	24	6.5	7.4	8.5	8.3	6.9	6.1	7.28
	25	6.4	5.9	5.4	6.9	6.5	6.2	6.22
	26	5.8	6.0	6.8	7.1	6.9	7.8	6.73
	27	8.5	7.9	9.4	9.8	10.6	11.4	9.60
	28	11.0	11.0	11.4	11.4	11.3	10.9	11.17
	29	10.6	5.0	3.6	3.9	4.0	3.6	5.10
	30	3.6	4.4	4.3	4.6	5.3	5.1	4.53
	31	5.8	6.5	8.1	7.8	7.8	7.6	7.27
	Mean	7.76	7.83	8.19	8.32	8.36	7.82	8.04

TABLE IIIb. ABSOLUTE HUMIDITY.

Month & Year	Day	2h am	6h am	10h am	2h pm	6h pm	10h pm	Mean
YAMAGUCHI 35m.								
IX 1889		mm	mm	mm	mm	mm	mm	mm
	28	13.6	12.5	15.5	13.4	12.6	13.1	13.45
	29	12.1	12.7	10.8	11.2	11.5	11.9	11.70
	30	12.0	7.8	8.0	7.7	9.2	9.7	9.07
	Mean	14.04	13.58	14.64	14.67	14.72	14.34	14.33
X 1889	1	8.3	7.6	9.4	8.8	7.9	8.7	8.45
	2	8.9	8.6	10.1	9.8	11.8	11.0	10.03
	3	11.0	9.7	10.5	9.6	11.5	10.2	10.42
	4	10.3	9.4	11.8	11.7	10.3	9.8	10.55
	5	9.4	9.5	11.0	10.5	12.9	14.0	11.22
	6	14.9	13.6	14.7	15.4	13.2	14.0	14.30
	7	11.0	10.2	14.1	12.7	11.5	11.9	11.90
	8	11.7	10.1	11.1	10.4	9.2	11.0	10.58
	9	11.1	10.0	10.2	9.5	10.2	10.5	10.25
	10	9.8	9.2	10.9	7.0	10.5	10.1	9.58
	11	8.9	8.7	10.7	9.0	10.7	10.6	9.77
	12	10.9	10.6	10.7	10.0	10.0	10.9	10.52
	13	8.9	8.4	8.4	5.7	10.0	8.8	8.37
	14	7.5	7.5	8.8	8.8	11.2	12.1	9.32
	15	12.0	12.5	14.9	14.3	14.5	14.4	13.77
	16	14.4	12.5	10.7	11.7	12.3	11.1	12.12
	17	9.0	8.7	12.3	10.5	11.8	11.7	10.67
	18	10.5	10.2	11.3	11.4	12.7	10.0	11.00
	19	9.2	9.2	11.9	11.2	12.3	9.0	10.47
	20	9.5	9.5	11.0	12.7	14.0	14.6	11.88
	21	17.2	16.6	13.8	9.7	8.3	7.3	12.15
	22	9.1	8.5	7.1	5.7	6.9	7.5	7.48
	23	6.5	5.9	8.2	6.9	7.7	7.8	7.17
	24	7.7	8.3	9.2	9.2	9.2	9.1	8.78
	25	7.3	6.6	9.7	7.3	8.6	8.4	8.08
	26	7.6	7.4	8.8	7.8	8.2	9.6	8.23
	27	9.3	9.3	10.1	9.7	12.3	11.9	10.43
	28	13.0	12.5	14.1	13.2	13.1	11.5	12.90
	29	8.0	6.0	4.4	4.3	4.7	5.4	5.47
	30	4.6	4.9	6.2	5.0	5.4	5.7	5.30
	31	6.4	7.1	8.3	8.9	9.5	8.7	8.15
	Mean	9.80	9.31	10.46	9.64	10.46	10.25	9.97

TABLE IV. RELATIVE HUMIDITY.

Month & Year	Day	2h am	4h am	6h am	7h am	8h am	9h am	10h am	11h am	Noon	1h pm	2h pm	3h pm	4h pm	5h pm	6h pm	8h pm	10h pm	Mid night	Mean
										FUJI 3718m.										
		%	%	%	%	%	%	%	%	%	%	%	%	%	%	%	%	%	%	%
VIII 1880	4	..	..	93	81	81	69	83	61	57	51	41	51	40	64	64	..	..	..	..
	5	..	..	76	71	94	89	89	90	90	81	83	84	85	85	88	..	..	..	..
IX 1887	4	..	..	..	..	..	..	..	..	..	..	..	..	89	69	66	92	84	100	..
	5	38	83	74	76	81	78	72	48	54	46	55	..	55	..	75	..	90	..	..
VIII 1889	1	65	64	60	..	50	..	58	..	39	..	33	..	43	..	59	69	62	68	55,1
	2	71	77	72	..	50	..	45	..	31	..	39	..	47	..	53	67	85	83	60,0
	3	84	84	50	..	44	..	37	..	33	..	43	..	56	..	55	60	67	78	57,6
	4	50	41	63	..	44	..	39	..	48	..	50	..	13	..	17	58	61	60	45,6
	5	30	39	20	..	31	..	48	..	22	..	41	..	50	..	67	64	77	77	47,2
	6	72	25	70	..	67	..	54	..	30	..	36	..	40	..	43	34	86	23	50,8
	7	27	27	27	..	24	..	18	..	16	..	43	..	54	..	45	31	30	85	31,4
	8	37	68	50	..	76	..	69	73	55	51	48	50	78	100	84	100	93	100	71,5
	9	100	100	94	90	94	70	64	71	69	66	85	55	93	92	90	96	84	93	88,5
	10	89	64	75	63	48	48	49	37	39	47	64	56	69	56	51	62	95	93	65,5
	11	100	97	89	82	92	61	48	47	24	28	36	46	58	70	73	84	79	77	71,4
	12	67	43	41	63	56	43	37	28	39	27	47	53	68	62	76	89	93	90	62,2
	13	100	100	100	100	94	72	66	54	51	51	51	88	80	75	79	87	95	93	83,0
	14	100	100	92	87	68	70	58	65	54	77	60	65	80	81	83	93	95	91	81,2
	15	81	63	57	52	36	38	32	32	28	35	35	30	24	24	14	23	39	59	40,9
	16	54	66	67	62	54	49	40	39	29	31	29	50	49	51	56	61	68	75	54,7
	17	77	63	63	46	44	47	39	34	45	41	37	39	57	64	64	100	100	100	65,7
	18	100	100	100	100	100	100	99	100	100	100	100	100	100	100	100	100	100	100	99,9
	19	100	100	100	100	100	100	100	100	100	100	100	100	100	100	100	100	100	100	100,0
	20	100	100	97	97	98	94	98	97	98	97	98	97	98	97	98	98	98	99	98,3
	21	96	98	93	97	98	97	98	97	98	83	93	87	86	89	85	74	69	75	88,1
	22	72	88	87	82	80	61	61	50	56	58	63	74	83	78	81	83	78	75	75,8
	23	83	93	84	77	69	54	51	49	56	63	70	65	73	88	100	96	97	97	80,7
	24	93	90	84	78	88	91	88	86	85	68	66	64	96	85	80	84	80	74	81,0
	25	89	88	80	81	75	73	68	64	62	61	69	73	83	87	89	87	98	82	81,3
	26	55	50	81	57	76	57	49	49	53	80	85	100	100	100	100	100	100	100	79,1
	27	100	100	100	100	100	99	100	100	100	98	100	100	99	100	100	100	82	100	98,4
	28	100	97	100	100	97	100	100	100	97	100	100	100	100	100	100	100	100	100	99,0
	29	100	100	100	100	100	100	100	100	100	100	100	100	100	100	100	100	100	100	100,0
	30	100	100	100	100	94	90	52	47	41	36	29	24	26	37	58	51	23	22	57,7
	31	22	20	21	43	27	50	22	65	10	37	22	24	23	25	40	41	47	72	32,3
	Mean	77,9	78,0	74,9	80,7	70,0	72,9	61,3	66,0	55,1	64,4	60,3	68,4	68,5	77,1	72,2	76,9	78,8	80,7	71,2
IX 1889	1	89	..	100	..	..	..	32	..	..	..	85	..	..	..	100	..	100	..	84,3
	2	100	..	100	..	..	..	100	..	..	..	100	..	..	..	100	..	100	..	100,0
	3	100	..	100	..	..	..	88	..	..	..	74	..	..	..	49	..	47	..	76,3
	4	86	..	98	..	..	..	70	..	..	..	23	..	..	..	42	..	34	..	58,8
	5	100	..	100	..	..	..	88	..	..	..	100	..	..	..	100	..	100	..	98,0
	6	100	..	100	..	..	..	100	..	..	..	100	..	..	..	85	..	36	..	86,8
	7	25	..	44	..	..	..	20	..	..	..	16	..	..	..	100	..	100	..	50,8
General Mean		79,3	78,0	78,0	80,7	70,0	72,9	63,1	66,0	55,1	64,4	62,3	68,4	68,5	77,4	74,0	76,9	78,0	80,7	72,7
										ONTAKE 3062m.										
VIII 1894	1	42	43	48	..	35	..	54	..	38	..	40	..	49	..	70	81	100	100	58,3
	2	100	100	100	..	100	..	100	..	100	..	100	..	100	..	100	100	100	100	100,0
	3	100	100	100	..	100	..	100	..	100	..	100	..	100	..	100	100	100	100	100,0
	4	100	100	100	..	100	..	100	..	100	..	100	..	100	..	100	82	91	100	97,7
	5	100	100	100	..	100	..	95	..	84	..	100	..	100	..	100	100	100	100	98,2
	6	100	100	100	..	100	..	100	..	100	..	100	..	100	..	100	98	98	100	99,7
	7	100	100	100	..	90	..	82	..	67	..	71	..	68	..	85	98	62	49	81,0
	8	47	86	77	..	70	..	55	..	60	..	69	..	53	..	70	66	63	66	64,4
	9	57	64	69	..	68	..	65	..	68	..	76	..	77	..	76	64	99	100	73,6
	10	100	100	100	..	100	..	100	..	97	..	93	..	100	..	90	100	100	100	99,1
	11	100	97	100	..	100	..	100	..	100	..	100	..	100	..	100	97	100	100	99,5
	12	97	94	76	..	84	..	71	..	61	..	68	..	66	..	92	89	100	100	82,5
	13	100	100	100	..	100	..	86	..	70	..	73	..	75	..	87	78	82	79	85,8
	14	74	82	90	..	73	..	66	..	61	..	64	..	72	..	89	90	99	63	77,7
	15	60	66	76	..	43	..	46	..	56	..	63	..	83	..	71	80	100	90	69,5

Month & Year	Day	2h am	4h am	6h am	7h am	8h am	9h am	10h am	11h am	Noon	1h pm	2h pm	3h pm	4h pm	5h pm	6h pm	7h pm	8h pm	9h pm	10h pm	11h pm	Mid-night	Mean
HARA 3m.																							
		%	%	%	%	%	%	%	%	%	%	%	%	%	%	%	%	%	%	%	%	%	%
VIII 1880	4	..	..	91	80	79	79	71	68	65	65	71	70	75	89	83	..	..	..	..	..	..	..
	5	..	..	93	90	81	81	74	67	61	69	94	84	77	71	88	..	..	..	..	..	..	..
TOKIO 21m.																							
IX 1887	4	..	..	..	..	..	..	..	..	..	..	..	..	85	84	87	91	92	90	91	92	95	..
	5	93	92	93	90	80	66	67	57	51	53	50	..	48	..	60	..	..	..	83	..	..	..
YAMANAKA 990m.																							
VIII 1889	1	94	..	91	94	84	73	79	66	69	72	64	73	78	84	85	93	94	94	95	..	..	84,8
	2	94	..	94	92	89	73	69	64	61	61	60	69	65	71	72	71	86	91	93	..	..	80,3
	3	95	..	98	92	85	80	69	66	66	65	61	64	74	76	78	88	92	92	94	..	..	82,5
	4	96	..	94	91	83	70	65	62	66	69	74	69	71	72	77	87	89	91	92	..	..	83,0
	5	94	..	96	87	81	81	74	63	60	66	70	75	76	77	82	89	91	94	92	..	..	84,7
	6	96	..	96	91	84	71	71	65	60	63	65	69	73	80	89	94	94	94	94	..	..	85,3
	7	94	..	94	82	79	71	70	61	61	60	67	75	79	86	91	92	92	94	94	..	..	85,0
	8	94	..	93	86	77	77	77	69	61	76	80	73	81	92	96	96	94	96	96	..	..	89,3
	9	96	..	94	98	86	84	79	78	73	69	75	77	84	84	89	92	94	96	96	91	96	88,2
	10	97	..	95	91	77	76	76	71	71	65	64	73	81	88	93	94	93	93	96	95	95	87,2
	11	96	..	94	92	81	83	75	73	71	72	75	79	85	90	91	91	93	97	96	96	97	87,8
	12	96	..	96	88	86	81	73	65	69	66	70	68	79	80	89	92	93	92	94	94	94	86,8
	13	95	..	94	88	79	73	69	65	60	68	70	75	78	78	84	92	92	94	92	92	92	84,0
	14	96	..	96	86	77	77	71	70	69	69	73	74	91	91	86	89	88	91	94	93	95	86,5
	15	95	..	95	91	73	69	66	64	56	64	61	62	71	72	81	87	87	93	93	91	92	82,3
	16	92	..	94	90	76	68	63	63	61	61	63	58	57	69	82	89	87	87	93	..	93	81,2
	17	94	..	93	87	69	69	66	59	60	59	61	58	74	83	88	91	91	91	91	..	..	82,2
	18	94	..	93	89	85	91	85	77	88	95	96	98	98	100	96	97	98	100	100	100	100	94,0
	19	100	..	100	100	100	99	98	93	100	98	98	98	96	96	99	96	96	97	96	..	98	98,5
	20	96	..	98	98	100	100	100	100	99	100	99	100	100	97	100	99	100	99	100	100	99	98,8
	21	99	..	99	98	96	95	89	88	87	85	85	91	88	92	87	94	98	98	97	..	97	92,7
	22	100	..	98	91	85	83	81	71	71	71	73	83	83	87	92	95	95	98	98	..	98	90,0
	23	100	..	97	95	88	82	74	69	65	67	71	82	92	97	98	95	98	98	98	99	98	89,2
	24	99	..	99	94	89	82	76	74	77	76	74	75	85	81	89	91	94	98	97	96	98	89,0
	25	98	..	96	94	82	82	81	74	76	80	81	100	88	90	92	93	94	96	96	95	96	90,7
	26	95	..	95	90	87	87	80	77	66	62	64	69	84	96	97	99	97	99	99	98	98	88,3
	27	97	..	99	98	99	99	99	98	98	98	94	98	96	96	99	100	98	98	99	99	99	97,8
	28	99	..	99	98	94	95	94	96	97	97	95	97	96	98	99	98	98	98	100	99	99	97,8
	29	98	..	99	99	98	98	98	98	98	98	98	98	98	97	98	99	98	99	96	99	96	97,8
	30	95	..	97	98	98	97	96	91	88	85	81	81	83	86	93	95	97	98	96	98	95	93,0
	31	96	..	98	95	95	90	84	82	81	82	83	79	86	90	93	93	96	93	91	93	94	90,5
	Mean	96,1	..	95,9	92,5	85,8	82,5	78,8	74,8	73,7	74,5	75,9	78,7	83,0	86,3	89,9	92,8	93,8	95,1	95,3	96,2	96,3	88,68
IX 1889	1	92	..	94	..	..	..	85	..	..	..	73	..	..	..	90	..	..	..	88	..	..	87,0
	2	94	..	87	..	..	..	98	..	..	..	94	..	..	..	98	..	..	..	95	..	..	94,3
	3	98	..	98	..	..	..	94	..	..	..	90	..	..	..	96	..	..	..	98	..	..	95,7
	4	98	..	98	..	..	..	86	..	..	..	85	..	..	..	94	..	..	..	96	..	..	92,8
	5	97	..	99	..	..	..	92	..	..	..	94	..	..	..	99	..	..	..	98	..	..	96,5
	6	98	..	95	..	..	..	82	..	..	..	86	..	..	..	95	..	..	..	96	..	..	92,2
	7	98	..	99	..	..	..	92	..	..	..	96	..	..	..	94	..	..	..	96	..	..	95,8
General Mean		96,2	..	96,0	92,5	85,8	82,5	80,8	74,8	73,7	74,6	78,2	78,7	83,0	86,3	90,0	92,8	93,8	95,1	95,3	96,2	96,3	89,7
KUROSAWA 832m.																							
VIII 1891	1	94	93	89	..	71	..	50	..	44	..	44	..	49	..	68	..	89	..	94	..	91	72,7
	2	98	90	95	..	82	..	69	..	82	..	84	..	82	..	95	..	98	..	98	..	98	90,0
	3	98	97	98	..	89	..	77	..	64	..	56	..	61	..	66	..	78	..	82	..	78	78,7
	4	78	81	98	..	87	..	87	..	89	..	68	..	86	..	92	..	98	..	93	..	93	87,5
	5	97	98	99	..	88	..	83	..	71	..	75	..	87	..	91	..	96	..	97	..	99	90,1
	6	98	98	98	..	91	..	74	..	75	..	76	..	88	..	77	..	96	..	99	..	97	88,9
	7	97	98	97	..	85	..	69	..	64	..	74	..	91	..	90	..	96	..	94	..	97	87,7
	8	95	98	98	..	79	..	72	..	50	..	55	..	43	..	58	..	86	..	93	..	94	76,7
	9	97	95	96	..	79	..	60	..	55	..	60	..	47	..	71	..	88	..	94	..	97	78,2
	10	98	97	97	..	82	..	71	..	62	..	55	..	65	..	78	..	92	..	99	..	99	82,9
	11	98	97	98	..	90	..	70	..	72	..	58	..	61	..	72	..	96	..	98	..	96	83,8
	12	98	99	98	..	78	..	50	..	60	..	55	..	52	..	71	..	98	..	92	..	96	79,2
	13	98	97	92	..	86	..	65	..	55	..	68	..	66	..	66	..	89	..	98	..	89	80,7
	14	96	99	99	..	89	..	64	..	50	..	55	..	56	..	72	..	80	..	90	..	98	78,2
	15	94	90	89	..	72	..	57	..	50	..	54	..	50	..	63	..	95	..	97	..	99	75,6

Month & Year	Day	2h am	4h am	6h am	7h am	8h am	9h am	10h am	11h am	Noon	1h pm	2h pm	3h pm	4h pm	5h pm	6h pm	8h pm	10h pm	Mid-night	Mean
ONTAKE 3062m.																				
		%	%	%	%	%	%	%	%	%	%	%	%	%	%	%	%	%	%	%
VIII 1891	16	98	100	97	..	97	..	100	..	95	..	100	..	100	..	100	100	100	100	98,5
	17	100	100	100	..	100	..	100	..	100	..	100	..	100	..	100	100	100	100	100,0
	18	100	100	100	..	100	..	100	..	100	..	100	..	100	..	96	89	100	83	97,8
	19	71	60	76	..	50	..	82	..	48	..	56	..	77	..	53	23	41	31	56,6
	20	32	23	16	..	34	..	30	..	32	..	49	..	61	..	40	21	14	29	31,7
	21	25	100	100	..	100	..	100	..	100	..	100	..	100	..	100	100	100	100	93,7
	22	100	100	100	..	100	..	100	..	100	..	100	..	100	..	100	100	100	100	100,0
	23	100	100	100	..	100	..	100	..	100	..	100	..	100	..	100	85	94	85	96,7
	24	84	87	88	..	83	..	81	..	80	..	81	..	81	..	86	91	93	85	85,0
	25	84	92	98	..	87	..	65	..	72	..	70	..	70	..	55	62	60	33	70,7
	26	26	41	31	..	41	..	43	..	65	..	68	..	70	..	48	44	17	16	42,5
	27	18	12	16	..	29	..	67	..	51	..	74	..	71	..	88	86	74	36	52,7
	28	56	53	84	..	94	..	61	..	73	..	81	..	82	..	80	56	50	99	72,4
	29	99	100	100	..	100	..	95	..	74	..	66	..	70	..	63	33	2	4	67,2
	30	2	50	25	..	20	..	35	..	45	..	50	..	63	..	53	73	7	6	36,5
	31	28	28	31	..	38	..	40	..	60	..	74	..	81	..	45	47	46	26	46,4
	Mean	74,0	80,2	80,9	..	78,8	..	78,0	..	76,0	..	79,7	..	83,0	..	82,1	78,4	77,2	73,8	78,5
IX 1891	1	79	80	81	..	74	..	63	..	53	..	60	..	70	..	98	95	99	100	78,9
	2	100	100	93	..	100	..	86	..	68	..	69	..	80	..	93	95	99	86	88,3
	3	92	85	79	..	72	..	56	..	73	..	73	..	73	..	83	81	69	74	75,8
	4	55	50	72	..	60	..	56	..	64	..	64	..	85	..	76	62	57	63	63,7
	5	70	54	62	..	83	..	55	..	41	..	44	..	62	..	69	75	54	50	59,9
	6	48	83	85	..	64	..	51	..	60	..	57	..	64	..	61	68	79	40	63,3
	7	33	30	41	..	48	..	27	..	56	..	56	..	47	..	45	47	55	84	47,4
	8	100	100	96	..	85	..	100	..	100	..	100	..	100	..	100	100	100	100	98,4
	9	100	100	100	..	100	..	100	..	100	..	100	..	100	..	100	100	100	100	100,0
	10	100	100	100	..	78	..	64	..	100	..	100	..	100	..	100	100	100	100	95,2
	11	100	100	100	..	100	..	100	..	95	..	100	..	100	..	100	97	100	100	99,3
	12	100	100	97	..	97	..	87	..	95	..	71	..	88	..	100	100	100	100	94,6
General Mean		76,1	80,7	81,7	..	79,2	..	75,9	..	75,9	..	78,3	..	82,5	..	82,9	80,3	79,0	76,4	79,0
GOZAISHODAKE 1200m.																				
IX 1888	4	100	100	100	..	100	..	100	..	100	..	100	..	100	..	100	100	100	100	100
	5	100	100	100	..	96	..	95	..	84	..	83	..	85	..	89	92	91	92	92
	6	93	93	88	..	81	..	65	..	58	..	72	..	82	..	86	80	100	100	83
	7	100	100	100	..	100	..	81	..	71	..	69	..	71	..	87	63	74	83	84
	8	90	100	100	..	100	..	97	..	96	..	91	..	81	..	70	83	82	77	88
	9	88	98	100	..	16	..	84	..	97	..	100	..	100	..	100	100	96	100	97
	10	100	100	100	..	100	..	100	..	100	..	100	..	100	..	100	100	100	100	100
	11	100	100	100	..	100	..	100	..	100	..	100	..	100	..	100	100	100	100	100
	12	100	100	100	..	100	..	100	..	100	..	100	..	100	..	100	100	100	100	100
	13	100	100	88	..	46	..	78	..	69	..	100	..	83	..	99	100	100	100	89
	14	100	100	100	..	100	..	82	..	86	..	83	..	89	..	91	89	86	89	90
	15	69	68	72	..	76	..	62	..	72	..	79	..	88	..	100	100	100	100	82
	16	100	100	100	..	100	..	100	..	100	..	98	..	90	..	97	98	91	100	98
	17	100	100	100	..	99	..	84	..	87	..	78	..	76	..	93	85	69	75	87
	18	91	100	100	..	99	..	74	..	62	..	72	..	77	..	99	91	84	68	84
	19	60	68	100	..	93	..	94	..	89	..	92	..	100	..	100	100	100	100	92
	20	100	100	100	..	100	..	100	..	100	..	100	..	100	..	100	100	100	100	100
	21	100	100	100	..	100	..	100	..	93	..	91	..	93	..	96	100	99	100	98
	22	100	95	100	..	100	..	97	..	100	..	100	..	100	..	100	100	100	100	99
	23	100	100	100	..	100	..	89	..	79	..	78	..	85	..	95	99	100	76	92
	24	52	32	33	..	46	..	59	..	72	..	94	..	100	..	100	98	95	92	71
	25	64	21	16	..	40	..	89	..	87	..	95	..	99	..	100	100	100	100	76
	26	100	100	100	..	100	..	85	..	88	..	93	..	95	..	90	99	93	92	94
	27	100	100	100	..	91	..	85	..	85	..	73	..	66	..	84	93	94	89	89
	28	25	32	57	..	44	..	34	..	57	..	70	..	77	..	77	76	82	85	60
	29	100	100	90	..	88	..	78	..	80	..	90	..	85	..	100	98	85	79	89
	30	84	81	82	..	92	..	83	..	72	..	70	..	78	..	100	79	78	88	81
X 1888	1	82	89	69	..	86	..	98	..	81	..	70	..	75	..	69	67	76	95	80
	2	100	98	94	..	70	..	71	..	66	..	68	..	77	..	87	97	91	88	83
	3	84	86	85	..	75	..	63	..	61	..	52	..	75	..	87	85	66	..	69
	Mean	89	89	89	..	87	..	84	..	83	..	85	..	87	..	83	92	91	89	85

Month & Year	Day	2h am	4h am	6h am	7h am	8h am	9h am	10h am	11h am	Noon	1h pm	2h pm	3h pm	4h pm	5h pm	6h pm	7h pm	8h pm	9h pm	10h pm	11h pm	Mid-night	Mean
KUROSAWA 832m.																							
		%	%	%	%	%	%	%	%	%	%	%	%	%	%	%	%	%	%	%	%	%	%
VIII 1891	16	99	99	98	..	81	..	48	..	46	..	45	..	53	..	71	..	89	..	86	..	89	75,3
	17	93	88	80	..	88	..	94	..	71	..	66	..	90	..	99	..	99	..	99	..	100	88,9
	18	99	100	99	..	97	..	98	..	66	..	98	..	96	..	93	..	98	..	100	..	95	94,1
	19	94	100	97	..	87	..	65	..	52	..	52	..	49	..	62	..	77	..	95	..	99	76,7
	20	99	99	97	..	81	..	54	..	44	..	53	..	60	..	99	..	100	..	100	..	99	82,2
	21	100	100	92	..	96	..	91	..	95	..	98	..	95	..	99	..	93	..	97	..	90	96,6
	22	99	95	99	..	92	..	91	..	91	..	86	..	81	..	91	..	98	..	98	..	99	93,6
	23	99	93	100	..	91	..	92	..	95	..	94	..	94	..	95	..	99	..	98	..	96	96,2
	24	94	96	95	..	88	..	78	..	73	..	72	..	77	..	90	..	95	..	96	..	98	87,7
	25	98	96	93	..	80	..	61	..	49	..	49	..	77	..	81	..	94	..	93	..	95	80,5
	26	92	94	94	..	85	..	66	..	49	..	41	..	46	..	82	..	83	..	91	..	97	76,7
	27	97	98	100	..	78	..	59	..	56	..	46	..	58	..	84	..	88	..	96	..	98	79,8
	28	97	97	99	..	87	..	68	..	54	..	49	..	50	..	77	..	95	..	96	..	97	80,5
	29	96	97	95	..	86	..	70	..	55	..	44	..	42	..	72	..	83	..	93	..	97	77,6
	30	97	100	97	..	81	..	61	..	39	..	39	..	60	..	78	..	84	..	94	..	97	77,2
	31	97	99	97	..	78	..	62	..	48	..	51	..	58	..	82	..	97	..	95	..	91	79,6
	Mean	96,4	96,5	95,9	..	84,5	..	70,2	..	62,1	..	61,6	..	65,8	..	80,0	..	91,8	..	94,9	..	95,7	83,0
IX 1891	1	96	96	97	..	80	..	62	..	60	..	51	..	53	..	78	..	96	..	95	..	95	79,7
	2	94	96	98	..	84	..	67	..	57	..	75	..	73	..	92	..	98	..	97	..	98	85,7
	3	93	98	95	..	81	..	64	..	47	..	59	..	56	..	84	..	93	..	96	..	95	80,1
	4	94	98	95	..	81	..	70	..	60	..	56	..	57	..	78	..	91	..	98	..	98	81,2
	5	99	98	96	..	82	..	55	..	44	..	51	..	63	..	76	..	95	..	94	..	99	79,3
	6	95	90	94	..	84	..	66	..	51	..	51	..	51	..	67	..	87	..	90	..	95	76,8
	7	96	97	97	..	78	..	62	..	56	..	58	..	65	..	78	..	87	..	96	..	95	80,4
	8	96	98	86	..	80	..	74	..	69	..	68	..	88	..	93	..	94	..	95	..	99	86,7
	9	98	100	99	..	98	..	96	..	87	..	87	..	87	..	96	..	94	..	100	..	99	95,1
	10	100	98	98	..	91	..	56	..	56	..	64	..	64	..	80	..	98	..	98	..	97	82,8
	11	98	100	100	..	84	..	62	..	50	..	54	..	53	..	88	..	93	..	97	..	99	81,2
	12	98	99	100	..	85	..	63	..	51	..	46	..	49	..	86	..	89	..	96	..	97	80,2
General Mean		96,4	96,8	96,0	..	84,3	..	69,2	..	60,8	..	61,0	..	65,7	..	80,9	..	92,1	..	95,2	..	96,1	82,9
YOKKAICHI 4m.																							
IX 1888	4	97	..	88	..	..	..	65	..	..	..	66	..	..	..	87	..	..	..	93	..	..	83
	5	93	..	87	..	..	..	75	..	..	..	55	..	..	..	64	..	..	..	72	..	..	74
	6	85	..	85	..	..	..	61	..	..	..	46	..	..	..	60	..	..	..	90	..	..	71
	7	94	..	96	..	..	..	69	..	..	..	59	..	..	..	74	..	..	..	91	..	..	80
	8	94	..	98	..	..	..	81	..	..	..	79	..	..	..	88	..	..	..	89	..	..	88
	9	92	..	92	..	..	..	75	..	..	..	81	..	..	..	88	..	..	..	91	..	..	87
	10	94	..	87	..	..	..	98	..	..	..	93	..	..	..	95	..	..	..	98	..	..	94
	11	96	..	95	..	..	..	96	..	..	..	94	..	..	..	79	..	..	..	94	..	..	92
	12	91	..	96	..	..	..	95	..	..	..	92	..	..	..	84	..	..	..	91	..	..	91
	13	86	..	92	..	..	..	65	..	..	..	68	..	..	..	40	..	..	..	85	..	..	73
	14	83	..	85	..	..	..	60	..	..	..	51	..	..	..	79	..	..	..	83	..	..	76
	15	85	..	96	..	..	..	64	..	..	..	55	..	..	..	70	..	..	..	91	..	..	76
	16	80	..	94	..	..	..	89	..	..	..	81	..	..	..	80	..	..	..	96	..	..	88
	17	98	..	96	..	..	..	61	..	..	..	56	..	..	..	72	..	..	..	86	..	..	78
	18	85	..	90	..	..	..	70	..	..	..	51	..	..	..	68	..	..	..	72	..	..	73
	19	86	..	92	..	..	..	66	..	..	..	66	..	..	..	80	..	..	..	87	..	..	79
	20	92	..	99	..	..	..	92	..	..	..	89	..	..	..	91	..	..	..	93	..	..	93
	21	98	..	95	..	..	..	97	..	..	..	98	..	..	..	91	..	..	..	97	..	..	95
	22	97	..	97	..	..	..	82	..	..	..	92	..	..	..	95	..	..	..	95	..	..	93
	23	87	..	94	..	..	..	65	..	..	..	60	..	..	..	75	..	..	..	86	..	..	78
	24	92	..	93	..	..	..	71	..	..	..	63	..	..	..	74	..	..	..	82	..	..	79
	25	83	..	77	..	..	..	64	..	..	..	76	..	..	..	69	..	..	..	90	..	..	76
	26	89	..	93	..	..	..	76	..	..	..	66	..	..	..	87	..	..	..	94	..	..	81
	27	96	..	97	..	..	..	83	..	..	..	72	..	..	..	84	..	..	..	95	..	..	88
	28	97	..	97	..	..	..	82	..	..	..	43	..	..	..	71	..	..	..	73	..	..	77
	29	83	..	90	..	..	..	69	..	..	..	61	..	..	..	79	..	..	..	89	..	..	78
	30	84	..	85	..	..	..	65	..	..	..	54	..	..	..	71	..	..	..	78	..	..	72
X 1888	1	82	..	85	..	..	..	67	..	..	..	63	..	..	..	67	..	..	..	61	..	..	71
	2	60	..	69	..	..	..	47	..	..	..	45	..	..	..	68	..	..	..	60	..	..	59
	3	62	..	86	..	..	..	45	..	..	..	29	..	..	..	68	..	..	..	81	..	..	63
	Mean	88	..	91	..	..	..	73	..	..	..	67	..	..	..	77	..	..	..	86	..	..	80

TABLE IV. RELATIVE HUMIDITY.

Month & Year	Day	2h am	6h am	10h am	2h pm	6h pm	10h pm	Mean
				HIGASHI HOBEN 736m.				
VIII 1889		%	%	%	%	%	%	%
	1	97	97	95	79	93	97	93,0
	2	100	97	87	98	91	100	95,5
	3	87	90	85	76	89	72	83,2
	4	65	71	72	76	77	61	70,3
	5	59	78	77	79	85	73	75,2
	6	95	100	88	79	88	100	91,7
	7	99	100	85	71	89	84	88,0
	8	100	98	82	73	79	55	81,2
	9	83	93	90	76	90	91	87,2
	10	84	81	87	87	99	65	83,3
	11	69	78	79	79	85	81	78,5
	12	98	91	89	69	82	74	82,2
	13	90	72	94	77	91	91	85,8
	14	94	100	88	74	95	93	90,7
	15	98	100	84	81	99	85	91,2
	16	99	94	99	84	85	93	92,8
	17	100	98	91	80	81	88	90,2
	18	89	87	77	78	81	91	83,8
	19	91	96	89	95	99	100	95,5
	20	97	96	99	98	98	100	98,0
	21	100	99	95	82	91	90	92,8
	22	84	98	99	87	99	97	94,0
	23	99	100	97	94	100	98	98,0
	24	99	100	100	94	99	98	98,3
	25	99	99	99	99	99	100	99,2
	26	99	99	99	99	97	100	98,8
	27	100	98	92	87	96	96	94,8
	28	99	84	77	83	88	91	87,0
	29	97	99	90	90	93	98	94,5
	30	77	93	85	93	93	95	89,3
	31	85	99	98	98	99	97	96,0
	Mean	91,03	93,06	89,29	84,35	91,39	88,93	89,68
IX 1889	1	97	99	96	95	99	100	97,7
	2	100	99	99	99	100	100	99,5
	3	96	88	74	76	88	86	84,7
	4	83	72	76	81	94	97	83,3
	5	98	99	99	93	98	99	97,7
	6	100	98	96	99	98	97	98,0
	7	96	99	84	80	98	94	91,8
	8	99	96	75	82	95	85	88,5
	9	89	72	64	78	91	99	80,7
	10	79	85	71	75	94	97	83,5
	11	100	100	99	99	100	100	99,7
	12	99	99	97	83	86	88	92,0
	13	89	85	82	81	91	65	82,2
	14	93	74	75	64	92	92	81,7
	15	96	86	72	88	87	89	86,3
	16	90	86	78	74	76	82	81,0
	17	78	78	72	69	83	84	77,3
	18	79	78	77	68	67	78	74,5
	19	73	93	74	76	92	99	84,5
	20	100	100	99	97	99	99	99,0
	21	99	100	99	99	93	94	97,3
	22	100	94	81	78	76	84	85,5
	23	89	84	80	86	96	91	87,7
	24	87	86	88	76	85	86	84,7
	25	90	82	73	76	100	100	87,8
	26	96	89	80	78	96	96	89,2
	27	99	99	99	100	100	99	99,3

TABLE IVa. RELATIVE HUMIDITY.

Month & Year	Day	2h am	6h am	10h am	2h pm	6h pm	10h pm	Mean
				YAMAGUCHI 35m.				
VIII 1889		%	%	%	%	%	%	%
	1	90	96	76	61	71	92	81,0
	2	93	96	75	72	63	91	81,7
	3	89	94	66	56	67	85	76,2
	4	91	95	64	66	76	89	80,2
	5	95	98	66	57	68	92	79,3
	6	94	92	76	59	68	92	80,2
	7	94	94	65	54	52	91	75,0
	8	95	93	68	50	64	89	76,5
	9	93	96	67	51	72	90	78,2
	10	94	95	74	54	72	88	79,5
	11	93	92	66	48	65	86	75,0
	12	91	92	74	57	61	89	77,3
	13	92	96	70	52	67	88	77,5
	14	90	85	68	48	70	70	71,8
	15	76	86	56	59	73	63	68,8
	16	80	79	70	61	62	71	70,5
	17	84	91	65	59	58	80	72,8
	18	90	93	59	51	61	67	70,2
	19	69	70	58	69	72	73	68,5
	20	84	90	84	65	74	91	81,3
	21	91	98	68	62	67	91	79,5
	22	90	92	69	69	78	92	80,2
	23	89	92	70	62	75	80	78,0
	24	80	82	68	63	73	80	74,3
	25	84	85	80	89	92	95	87,5
	26	85	81	64	65	72	76	73,8
	27	79	80	65	75	68	80	74,5
	28	78	78	54	54	72	74	67,8
	29	77	91	58	63	67	76	72,0
	30	82	83	65	68	70	89	75,8
	31	93	86	82	88	94	94	89,5
	Mean	87,26	89,39	68,00	61,23	69,81	83,90	76,60
IX 1889	1	92	94	77	66	75	82	81,0
	2	83	89	94	68	79	86	83,2
	3	91	87	62	59	71	91	76,8
	4	92	94	63	55	66	92	77,0
	5	93	94	75	70	74	85	81,7
	6	95	95	69	69	72	84	80,7
	7	93	95	64	55	71	89	77,8
	8	92	95	64	67	72	74	77,3
	9	83	93	53	50	69	77	70,8
	10	81	88	57	58	69	90	73,8
	11	83	80	73	68	79	80	77,2
	12	82	89	68	57	68	76	73,3
	13	93	93	63	56	64	85	75,7
	14	92	95	49	37	64	92	71,5
	15	92	98	67	65	72	89	80,5
	16	96	96	56	46	64	76	72,3
	17	91	94	53	45	67	84	72,3
	18	91	95	59	45	62	87	73,2
	19	93	93	59	52	89	90	79,3
	20	94	90	82	86	84	90	87,7
	21	94	95	92	94	84	91	91,5
	22	96	94	77	61	74	88	81,7
	23	91	96	70	67	84	92	83,3
	24	95	94	87	63	72	91	83,7
	25	90	95	84	63	76	91	83,2
	26	90	98	67	60	81	93	81,5
	27	95	96	88	94	94	94	93,0

TABLE IV. RELATIVE HUMIDITY.

Month & Year	Day	2^h am	6^h am	10^h am	2^h pm	6^h pm	10^h pm	Mean
HIGASHI HOBEN 736^m.								
IX 1889		%	%	%	%	%	%	%
	28	99	98	97	83	95	99	95.2
	29	100	100	87	75	89	79	88.3
	30	74	72	62	62	76	76	70.3
	Mean	91.93	89.63	83.70	82.33	91.03	91.13	88.30
X 1889	1	75	72	64	62	69	65	67.8
	2	38	53	61	56	86	83	62.8
	3	71	77	70	67	83	68	72.7
	4	83	87	71	62	65	74	74.2
	5	12	46	58	53	86	82	56.2
	6	100	100	100	100	100	100	100.0
	7	100	96	86	77	95	86	90.0
	8	89	89	67	67	73	57	73.7
	9	76	67	59	60	66	42	61.7
	10	73	93	51	51	84	47	66.5
	11	40	60	53	40	55	63	51.8
	12	57	68	59	67	58	78	62.8
	13	71	69	62	31	70	47	58.3
	14	54	42	69	64	75	99	67.2
	15	96	100	100	100	97	100	98.8
	16	100	100	79	90	96	90	92.5
	17	94	94	75	68	84	57	78.7
	18	42	88	63	56	91	75	69.2
	19	78	52	56	63	35	59	56.7
	20	71	72	73	85	99	100	83.3
	21	99	99	98	85	81	67	88.2
	22	77	78	69	65	68	69	71.0
	23	60	67	63	60	70	72	65.3
	24	78	89	86	76	79	72	80.0
	25	78	71	53	54	71	70	66.2
	26	67	69	66	57	71	80	68.3
	27	87	81	75	66	94	100	83.8
	28	99	100	100	100	99	100	99.7
	29	99	67	61	72	79	68	74.3
	30	68	88	76	59	81	81	76.7
	31	85	94	83	85	94	93	89.0
	Mean	74.74	78.32	71.16	67.35	79.89	75.61	74.43

TABLE IV^a. RELATIVE HUMIDITY.

Month & Year	Day	2^h am	6^h am	10^h am	2^h pm	6^h pm	10^h pm	Mean
YAMAGUCHI 35^m.								
IX 1889		%	%	%	%	%	%	%
	28	93	90	89	60	71	83	81.0
	29	82	92	61	56	77	88	76.0
	30	90	57	48	43	72	90	66.7
	Mean	90.60	91.70	69.00	61.07	73.83	86.73	78.82
X 1889	1	90	97	56	45	56	82	71.0
	2	88	91	73	53	77	91	78.8
	3	91	94	56	42	80	90	75.5
	4	93	95	59	48	63	89	74.5
	5	94	95	59	45	75	85	75.5
	6	93	93	87	82	78	89	87.0
	7	88	96	74	58	70	88	79.0
	8	91	98	58	50	56	84	72.8
	9	86	72	57	47	65	90	69.5
	10	95	96	61	29	63	88	72.0
	11	94	95	58	40	71	82	73.3
	12	86	85	58	44	59	78	68.3
	13	87	93	45	25	64	86	66.7
	14	87	92	60	48	76	93	76.0
	15	91	94	93	79	89	94	90.0
	16	93	78	58	60	78	93	76.7
	17	96	98	74	51	80	91	81.7
	18	91	90	58	57	83	89	78.0
	19	94	96	74	59	82	79	80.7
	20	87	88	53	63	83	81	75.8
	21	88	91	80	62	64	63	74.7
	22	88	89	54	44	68	89	72.0
	23	92	95	65	45	76	89	77.0
	24	87	93	81	59	76	91	81.2
	25	95	92	69	45	68	90	76.5
	26	93	95	65	40	66	95	75.7
	27	95	98	69	46	85	80	78.8
	28	89	86	96	76	82	94	87.2
	29	66	52	43	49	58	87	59.2
	30	91	93	67	42	65	90	74.7
	31	89	90	80	67	87	94	84.5
	Mean	89.93	90.64	65.81	51.61	72.35	87.23	76.26

Month & Year	Day	2h a.m.		4h a.m.		6h a.m.		7h a.m.		8h a.m.		9h a.m.		10h a.m.		11h a.m.		Noon	
																			FUJI
			mps		mps		mps		mps		mps		mps		mps		mps		mps
VIII 1889	1	W	7.1	W	5.9	W	3.8	..	..	NW	2.6	..	..	W	2.9	..	..	S	4.6
	2	WSW	8.3	WSW	8.3	WNW	5.2			W	2.5			NW	1.5		..	NE	2.1
	3	NW	4.6	NNW	4.4	NNE	3.3	..		NE	3.7			NW	4.2	..	..	NW	6.9
	4	SW	9.4	W	6.2	N	3.6		..	N	3.1			NNE	3.5		..	NW	4.5
	5	W	6.2	W	4.7	W	3.3			NNE	1.3		..	NNE	1.6			SSW	3.2
	6	SSW	17.1	SW	13.7	SW	11.6	..		W	5.1			SW	12.2	..		WNW	2.8
	7	SW	16.0	SSW	15.9	SW	17.7			SW	19.2			SW	16.2	..		WSW	10.3
	8	SW	14.9	SSW	20.2	SW	15.7			SW	19.5			S	3.7	NE	3.2	SW	9.4
	9	SW	16.5	SW	16.9	WSW	9.2	SW	3.2	W	5.8	WSW	3.2	WSW	2.5	WSW	1.7	S	1.9
	10	SW	8.0	SW	9.2	SW	4.2	SW	3.7	W	1.5	NNW	2.1	W	2.5	W	2.4	WSW	4.9
	11	S	11.6	SW	22.7	WSW	22.7	WSW	13.1	WSW	21.4	SW	21.0	WSW	20.2	W	13.6	SW	5.5
	12	W	3.7	NW	0.6	N	1.7	W	1.4	N	1.2	NW	1.6	NW	1.7	WSW	3.1	N	1.6
	13	W	4.5	W	11.4	SW	9.4	WSW	8.3	NW	2.9	NW	3.2	WSW	3.4	W	3.2	SW	4.9
	14	SW	7.2	SW	5.2	W	1.9	W	2.2	NNW	2.4	NNE	2.5	ENE	2.9	NNE	2.4	NE	3.2
	15	NE	0.5	—	0.0	—	0.0	WSW	1.5	ENE	3.0	E	2.4	E	4.4	E	3.1	SE	3.4
	16	WSW	6.2	WSW	4.3	WSW	4.6	SW	2.3	S	2.8	SSW	2.6	W	2.8	WSW	2.7	E	3.0
	17	S	3.7	S	2.2	—	0.0	ENE	1.7	SW	3.7	ENE	2.5	ENE	2.2	E	2.7	S	4.0
	18	SSE	17.4	SE	6.8	SSE	7.7	SSE	6.5	SSE	9.4	SSE	12.0	SSE	12.0	SSE	11.8	SSE	15.1
	19	S	6	SSW	5	SSE	6	S	5	SSE	6	S	5	SSE	6	SSW	5	S	6
	20	SW	6	SW	6	SW	6	SW	5	SW	4	SW	5	SW	6	SW	6	SW	3
	21	SW	6	SW	5	SW	5	SSW	5	SW	4	SW	4	SW	4	SW	4	SW	3
	22	W	2	W	2	W	2	W	2	W	2	—	0	W	1	W	1	W	1
	23	W	2	W	1	W	1	W	1	W	1	—	0	—	0	—	0	—	0
	24	—	0	—	0	—	0	—	0	—	0	—	0	—	0	—	0	—	0
	25	SW	8.1	SSW	10.7	SW	15.2	SW	9.3	SW	5.2	SW	5.6	SW	8.0	WSW	8.1	SW	14.4
	26	SW	28.8	SW	22.5	SW	24.7	SW	25.6	SW	25.8	WSW	19.6	SW	23.2	SW	22.4	SW	21.6
	27	WSW	4	WSW	3	WSW	1	—	0	W	1	—	0	—	0	SE	0.7	SE	3.2
	28	SSW	2	WSW	3	SW	1	SW	1	SW	2	W	1	W	2	SW	4.6	SW	3
	29	W	8.6	W	2.7	SW	4.2	SW	3.2	—	0.0	—	0.0	—	0.0	—	0.0	—	0.0
	30	W	2.4	W	2.7	NNW	3.2	NNW	5.2	NW	8.9	WSW	5.8	NW	6.7	W	6.6	NW	7.2
	31	W	5.2	W	4.2	W	5.0	W	2.2	NW	0.7	W	1.1	W	2.2	WSW	6.4	W	7.2
IX 1889	1	..	..	..	..	WSW	8.1	..	..	..	..	..	..	..	..	..	..	..	..
	2	SW	32.5	..	..	SW	34.3	..	..	..	..	..	..	SW	22.6	..	..	..	..
	3	SW	..	..	..	SW	..	..	..	..	..	..	..	SW	11.6	..	..	..	..
	4	SSW	14.2	..	..	SW	10.3	..	..	..	..	..	..	WSW	6.1	..	..	..	..
	5	W	9.3	..	..	NW	12.6	..	..	..	..	..	..	W	16.9	..	..	..	..
	6	SSW	..	..	..	SW	..	..	..	..	..	..	..	S	0.6	..	..	..	..
	7	WSW	..	..	..	W	..	..	..	..	..	..	..	SW	..	..	..	..	..
																			ONTAKE
VIII 1891	1	S	17.5	NNW	13.2	NW	12.2	..	..	NW	9.7	..	..	W	7.2	..	..	WNW	2.5
	2	W	14.5	SW	15.5	SW	12.7	..	..	WNW	12.2	..	..	NNW	10.7	..	..	WSW	13.2
	3	W	11.0	WSW	11.4	W	14.5	..	..	SW	18.7	..	..	SW	21.2	..	..	SSW	19.7
	4	SSW	23.6	SSW	16.1	WSW	11.2	..	..	SW	10.2	..	..	SW	13.8	..	..	SW	12.6
	5	W	10.7	W	7.7	W	3.7	..	..	SW	5.9	..	..	W	7.7	..	..	SW	2.2
	6	SW	10.2	W	4.0	SW	7.5	..	..	SW	6.5	..	..	SSE	2.9	..	..	SSW	4.9
	7	SW	11.1	SW	12.7	W	11.1	..	..	W	4.0	..	..	SW	4.2	..	..	W	4.2
	8	WNW	14.1	NW	15.7	N	6.8	..	..	N	4.7	..	..	NNW	6.7	..	..	N	6.8
	9	NW	7.4	WSW	14.0	W	10.7	..	..	W	7.8	..	..	SSW	2.6	..	..	W	2.6
	10	SW	15.5	SW	13.9	WSW	10.6	..	..	WSW	12.2	..	..	W	13.7	..	..	NW	8.4
	11	W	16.7	W	18.6	WNW	23.3	..	..	WSW	23.2	..	..	W	17.5	..	..	WSW	13.8
	12	WSW	17.1	WSW	22.2	WSW	17.2	..	..	WSW	14.3	..	..	WSW	7.4	..	..	W	6.6
	13	NW	17.3	NW	19.2	NW	23.7	..	..	NW	22.3	..	..	NW	12.5	..	..	WNW	4.9
	14	SW	13.2	W	20.7	W	18.2	..	..	WNW	11.2	..	..	W	7.4	..	..	WSW	4.8
	15	SW	8.0	W	7.0	WSW	8.9	..	..	SW	11.1	..	..	SW	12.2	..	..	SW	8.7
	16	S	18.2	S	16.4	S	25.6	..	..	S	23.0	..	..	S	20.3	..	..	S	18.8
	17	SW	30.3	SW	25.0	W	22.1	..	..	W	24.6	..	..	WNW	25.7	..	..	WNW	23.5
	18	W	14.3	W	11.6	WSW	18.1	..	..	WSW	16.5	..	..	WSW	21.7	..	..	WSW	22.3
	19	WNW	22.5	WNW	22.0	WNW	23.2	..	..	NW	14.4	..	..	WNW	17.8	..	..	NW	5.2

1^h pm		2^h pm		3^h pm		4^h pm		5^h pm		6^h pm		8^h pm		10^h pm		Midnight		Mean
3718m.																		
	mps		mps		mps		mps		mps		mps		mps		mps		mps	mps
..	..	W	5.0		..	W	4.2	..	..	W	3.2	W	4.2	W	3.3	W	8.2	4.58
..	..	NE	1.0	..	..	NNW	2.2	..	..	NNW	3.5	NW	2.2	W	2.5	NW	2.5	3.48
..	..	NE	5.9	..	..	NW	6.8	..	..	NW	9.5	NW	7.1	SW	5.8	SW	9.5	5.97
..	..	NW	2.9	..	..	W	6.9	..	..	NW	2.9	NW	3.3	NNW	1.7	NW	2.9	4.31
..	..	SSW	5.2	..	..	SSW	6.2	..	..	NW	5.2	W	7.3	W	11.7	SW	18.8	6.22
..	..	WSW	4.7	..	..	SW	2.6	..	..	NW	2.7	NW	12.4	SW	21.1	SSW	15.0	10.08
..	..	W	3.8	..	..	SW	10.7	..	..	SW	9.6	SW	9.9	SW	11.2	SW	7.0	12.29
SW	3.2	SSW	5.2	WSW	3.2	SW	9.9	SSW	8.1	SW	14.5	SW	9.2	WSW	20.5	SW	12.5	12.93
SW	2.0	SW	2.6	SSW	7.1	WSW	10.3	SW	14.8	WSW	9.3	W	3.1	WSW	4.1	SW	6.9	7.42
WSW	4.4	SW	5.7	WSW	6.4	SW	5.1	WSW	3.9	WSW	6.2	WSW	11.7	WSW	9.5	SW	3.0	6.12
SW	3.6	WSW	5.3	WSW	2.5	WSW	7.1	WNW	5.7	W	2.8	W	2.4	WSW	3.9	W	1.8	10.57
N	2.6	N	3.5	W	3.2	W	3.1	WNW	2.6	NW	2.9	N	2.2	N	3.7	NW	4.9	2.57
W	3.6	W	3.7	WNW	3.2	NW	2.6	WSW	3.6	WSW	2.3	NW	0.9	N	2.4	NW	4.3	4.39
NNE	2.5	ENE	1.5	W	1.2	NW	2.0	NE	0.7	—	0.0	NE	1.0	ENE	0.9	NE	1.3	2.46
NE	2.7	SW	2.2	NE	2.0	WSW	3.3	W	2.8	WSW	4.7	SW	3.5	SW	4.2	SW	6.2	2.95
W	3.4	W	1.4	ENE	2.9	W	2.8	NW	1.4	S	3.3	S	4.8	S	5.3	S	6.7	4.00
ENE	4.0	E	2.8	WSW	3.0	SW	2.6	NW	3.7	S	3.6	SSE	6.0	SSE	8.5	SSE	5.2	4.62
SSE	15.5	SSE	15.1	SSE	11.5	SSE	18.1	SSE	15.2	SSE	15.2	SSE	**6**	S	**6**	S	**6**	..
SW	**5**	SSW	**5**	SSW	**6**	SSW	**6**	SSW	**6**	SSW	**6**	SW	**6**	SW	**6**	SW	**6**	
SW	**6**	SW	**6**	SW	**6**	SW	**6**	SW	**3**	SW	**6**	SW	**6**	SW	**6**	SW	**5**	
SW	**3**	SW	**3**	WSW	**3**	WSW	**3**	W	**2**	W	**1**	—	**0**	W	**1**	W	**2**	
W	**1**	W	**1**	W	**1**	W	**1**	W	**1**	W	**2**	W	**2**	W	**3**	W	**3**	..
—	**0**	—	**0**	—	**0**	W	**1**	—	**0**	SW	**2**	—	**0**	W	**2**	—	**0**	
—	**0**	SW	**1**	S	**1**	—	**0**	—	**0**	SW	**1**	SW	2.9	SW	5.5	SW	5.2	
SW	2.7	WSW	8.9	W	6.7	WSW	10.6	SW	8.1	WSW	12.7	SW	19.0	SW	20.7	WSW	21.8	12.91
SW	21.6	SW	24.9	SW	22.1	SW	25.7	SW	**5**	SW	8.8	WSW	12.0	SW	28.1	SW	**5**	..
SW	3.5	SSE	3.4	SSW	2.4	SW	3.2	SSE	0.8	WSW	1.5	WSW	1.8	WSW	**3**	WSW	**3**	..
SSW	8.7	SSW	**2**	SSW	3.8	SSW	3.6	SSW	3.0	SSW	9.9	W	7.6	W	11.6	W	11.8	..
NW	0.5	NNW	1.2	NW	0.6	—	0.0	NW	1.8	W	0.9	SW	0.9	W	2.9	NW	0.9	1.86
WSW	6.1	NW	6.9	WSW	7.1	NW	6.2	W	6.4	NW	5.3	W	3.2	W	4.2	NW	7.0	5.24
SW	4.3	W	3.7	WSW	1.2	W	1.8	WSW	2.2	W	1.2	WSW	2.6	WSW	0.7	WSW	12.3	3.90
..	..	WSW	12.6	..	..	..	..	..	..	NNW	8.1	..	..	W	16.2	..	..	..
..	..	SW	..	..	..	..	..	..	..	SW	..	..	..	SW	..	..	..	..
..	..	SW	7.4	..	..	..	..	..	..	W	8.7	..	..	NW	9.3	..	..	..
..	..	W	8.2	..	..	..	..	..	..	W	12.4	..	..	W	13.2	..	..	..
..	..	NE	4.5	..	..	..	..	..	..	SSW	27.0	..	..	SSW	36.7	..	..	..
..	..	NW	..	..	..	..	..	..	..	WNW	..	..	..	W	..	..	..	..
..	..	S	..	..	..	..	..	..	..	SSW	..	..	..	SSW	..	..	..	..

Black figures refer to estimated forces.

3062m.

1^h pm		2^h pm		3^h pm		4^h pm		5^h pm		6^h pm		8^h pm		10^h pm		Midnight		Mean
..	..	W	1.4	..	..	W	5.5	..	..	W	10.2	W	11.2	W	10.3	WSW	20.3	10.10
..	..	W	12.0	..	..	W	18.5	..	..	NW	18.2	W	16.2	NW	12.3	NW	11.5	13.96
..	..	SSW	15.6	..	..	SSW	16.7	..	..	SW	32.7	SSW	35.2	SSW	37.3	SSW	31.7	22.14
..	..	SW	11.1	..	..	SW	13.0	..	..	W	14.6	WSW	21.0	W	15.9	WSW	9.7	15.23
..	..	SW	5.1	..	..	SW	8.7	..	..	W	6.7	SW	8.0	W	13.9	SW	13.2	7.79
..	..	WSW	5.6	..	..	SW	2.8	..	..	W	8.7	WSW	3.7	W	8.6	SW	10.6	6.32
..	..	N	2.0	..	..	N	2.2	..	..	WNW	7.2	W	6.4	NW	8.9	W	11.9	7.16
..	..	NNW	8.1	..	..	NW	4.9	..	..	N	1.7	N	2.3	N	3.3	N	6.0	6.76
..	..	NW	2.3	..	..	NW	2.7	..	..	SW	3.8	W	7.2	WSW	11.1	SW	12.9	7.09
..	..	WSW	8.8	..	..	WSW	6.5	..	..	W	11.3	NNW	11.6	WSW	11.2	W	15.7	11.62
..	..	W	11.0	..	..	WNW	11.2	..	..	WNW	13.8	WNW	15.7	W	12.1	W	18.7	16.30
..	..	WNW	4.9	..	..	WNW	11.0	..	..	WNW	14.3	W	16.4	WSW	17.8	WSW	17.0	13.85
..	..	NW	5.7	..	..	N	2.7	..	..	NW	5.6	NW	9.6	W	9.6	W	13.4	12.29
..	..	N	2.3	..	..	W	3.3	..	..	WNW	7.1	W	11.5	W	10.7	WSW	10.6	10.08
..	..	SSW	6.1	..	..	SSW	8.2	..	..	SSW	12.5	SW	13.9	SW	15.1	SW	16.5	10.68
..	..	S	31.7	..	..	SSE	27.3	..	..	S	22.7	S	23.0	S	27.1	SW	33.5	23.97
..	..	W	22.6	..	..	W	15.2	..	..	W	13.2	W	11.7	W	6.6	W	13.7	19.32
..	..	WSW	16.0	..	..	WSW	9.8	..	..	WSW	16.8	W	17.6	W	25.7	WNW	23.3	17.81
..	..	WNW	4.2	..	..	WNW	6.1	..	..	W	4.2	W	14.3	WNW	14.4	WNW	16.6	13.74

Month & Year	Day	2h a.m.	mps	4h a.m.	mps	6h a.m.	mps	7h a.m.	mps	8h a.m.	mps	9h a.m.	mps	10h a.m.	mps	11h a.m.	mps	Noon	mps
																		ONTAKE	
VIII 1891	20	W	12,2	WNW	16,1	WNW	18,2	..	..	WNW	15,1	..	..	W	6,5	..	..	WNW	6,5
	21	SW	12,0	SW	16,4	SW	13,0	..	..	SW	12,1	..	..	W	16,2	..	..	W	14,4
	22	WSW	9,4	WNW	24,7	W	20,7	..	..	W	10,4	..	..	W	6,2	..	..	WSW	8,5
	23	W	17,7	W	17,1	WSW	12,4	..	..	W	9,6	..	..	WSW	4,8	..	..	WSW	8,4
	24	W	15,3	WNW	11,6	W	13,7	..	..	WNW	13,4	..	..	SSW	6,2	..	..	W	7,2
	25	W	15,3	W	14,6	WNW	14,4	..	..	WNW	14,6	..	..	WNW	8,1	..	..	SW	4,1
	26	NW	18,1	WNW	20,7	NNW	8,6	..	..	SSW	3,2	..	..	NE	2,0	..	..	W	0,8
	27	WNW	8,2	NNW	8,7	NW	8,4	..	..	NW	1,2	..	..	WNW	1,8	..	..	SSE	2,2
	28	WSW	14,2	W	16,5	W	18,2	..	..	W	10,4	..	..	W	4,5	..	..	WSW	3,7
	29	WNW	20,5	WNW	21,9	NW	17,7	..	..	NW	18,7	..	..	NNW	11,1	..	..	N	7,4
	30	NW	16,5	W	11,9	N	9,7	..	..	W	4,4	..	..	N	1,4	..	..	NNW	1,4
	31	NW	17,2	NW	14,2	NW	12,2	..	..	ENE	2,2	..	..	ENE	1,9	..	..	ESE	1,9
	Mean		15,48		15,52		14,47		..		11,86		..		9,82		..		8,14
IX 1891	1	W	9,2	W	12,2	WNW	16,7	..	..	WNW	13,7	..	..	W	4,7	..	..	NNE	1,9
	2	W	14,9	W	11,6	W	13,2	..	..	NNW	10,2	..	..	WNW	7,7	..	..	NNE	2,0
	3	WNW	8,2	W	16,2	WNW	13,8	..	..	W	6,7	..	..	W	2,1	..	..	W	2,7
	4	W	6,8	SW	5,2	WSW	5,9	..	..	WSW	7,3	..	..	SW	4,9	..	..	NNW	3,4
	5	W	17,3	W	21,3	WNW	23,6	..	..	NNW	18,7	..	..	NW	15,7	..	..	W	3,1
	6	W	7,2	W	10,1	NW	7,7	..	..	NNW	0,9	..	..	NE	0,6	..	..	W	2,0
	7	WSW	12,3	W	10,3	WSW	6,8	..	..	WSW	7,0	..	..	WNW	6,5	..	..	NW	7,1
	8	SW	9,1	SW	12,6	WSW	12,2	..	..	WSW	8,7	..	..	SW	7,8	..	..	W	7,7
	9	SW	17,9	W	12,7	W	13,5	..	..	W	12,7	..	..	W	14,4	..	..	W	13,2
	10	SSW	16,6	SSW	25,2	SSW	25,2	..	..	SSW	29,0	..	..	SSW	26,7	..	..	SW	18,4
	11	SW	11,8	SW	12,0	SW	10,8	..	..	SW	10,6	..	..	SSW	14,3	..	..	SW	13,3
	12	SSW	18,6	S	17,7	S	16,8	..	..	SSW	14,1	..	..	SSW	8,7	..	..	SSE	6,3
General Mean			14,64		15,08		14,28		..		11,80		..		9,73		..		7,75
																		GOZAISHO-	
IX 1888	4	WSW	5,3	W	14,2	W	13,7	..	..	W	10,2	..	..	W	12,0	..	..	W	16,1
	5	NW	8,0	NNW	16,8	NW	8,7	..	..	NW	8,9	..	..	NW	8,5	..	..	NW	4,6
	6	NW	10,4	N	11,8	NNW	12,2	..	..	NW	8,7	..	..	N	1,6	..	..	S	1,6
	7	SE	10,0	S	8,5	S	9,0	..	..	S	5,8	..	..	SW	1,9	..	..	WSW	1,5
	8	WSW	2,6	WNW	4,9	W	3,3	..	..	NW	2,3	..	..	N	0,2	..	..	WNW	2,5
	9	N	3,6	NE	8,8	NE	4,8	..	..	NE	3,4	..	..	E	3,1	..	..	NE	5,0
	10	ESE	9,7	ESE	10,4	ESE	7,7	..	..	E	6,3	..	..	ESE	4,5	..	..	SE	7,9
	11	ESE	16,3	SE	12,8	SE	15,0	..	..	SE	14,2	..	..	SE	12,3	..	..	SSE	18,2
	12	SSE	20,1	S	19,3	SSW	19,6	..	..	SW	13,3	..	..	WNW	8,3	..	..	WNW	9,2
	13	NNW	7,8	N	8,2	NNW	9,4	..	..	NW	4,1	..	..	NW	0,8	..	..	W	1,5
	14	NW	12,9	NW	10,4	NW	7,1	..	..	NW	9,2	..	..	NNW	7,1	..	..	NNW	5,2
	15	SW	2,8	SE	2,0	SSE	3,3	..	..	SSE	3,5	..	..	SSE	4,5	..	..	SSE	2,5
	16	SE	10,3	ESE	10,5	ESE	7,8	..	..	SE	8,4	..	..	SE	3,2	..	..	SSW	1,6
	17	NNW	10,6	NNW	7,2	NNW	10,1	..	..	NW	11,8	..	..	NNW	7,2	..	..	NNW	12,2
	18	NNW	13,1	NNW	12,2	NNW	8,0	..	..	NNW	9,5	..	..	NNW	7,5	..	..	NNW	8,4
	19	E	3,9	E	2,8	SE	4,6	..	..	SE	5,1	..	..	SE	9,0	..	..	SE	7,0
	20	SE	11,7	SE	17,4	SSE	15,6	..	..	SSE	19,2	..	..	SSE	13,4	..	..	SSE	8,3
	21	SW	6,6	W	2,4	W	2,2	..	..	SW	2,6	..	..	WSW	3,4	..	..	WSW	1,4
	22	NNW	1,8	SSW	1,1	N	5,8	..	..	N	5,5	..	..	N	1,5	..	..	NNW	2,3
	23	N	12,8	NNW	13,2	NNW	13,7	..	..	NNW	11,4	..	..	NNW	5,6	..	..	NNW	4,5
	24	NW	7,8	NW	9,6	NW	9,7	..	..	NW	9,7	..	..	NW	9,9	..	..	NW	8,0
	25	NW	4,7	N	2,6	NE	4,3	..	..	SE	5,1	..	..	SSE	8,0	..	..	SSE	5,0
	26	SSE	7,8	SSE	3,8	SSE	5,3	..	..	S	3,7	..	..	S	0,8	..	..	SE	2,0
	27	SE	2,1	SE	1,9	WNW	3,1	..	..	NNW	1,3	..	..	S	0,4	..	..	NW	1,1
	28	NNW	8,1	NNW	6,6	NNW	6,6	..	..	NW	5,5	..	..	NW	4,8	..	..	NW	4,1
	29	WNW	10,7	WNW	9,4	NW	7,0	..	..	NW	2,8	..	..	NW	1,2	..	..	W	6,1
	30	N	10,0	NNW	10,4	NNW	8,6	..	..	NNW	3,6	..	..	NW	0,9	..	..	NE	0,2
X 1888	1	NNW	5,9	N	5,8	N	5,1	..	..	N	4,2	..	..	NW	6,3	..	..	NW	4,9
	2	NW	7,8	NW	20,0	NW	17,1	..	..	NNW	16,2	..	..	NNW	15,4	..	..	NW	14,5
	3	NNW	12,6	NNW	9,2	NNW	8,6	..	..	NNW	5,3	..	..	NNW	10,2	..	..	NNW	8,6
	Mean		8,6		9,1		8,6		..		7,4		..		5,8		..		5,9

1h pm		2h pm		3h pm		4h pm		5h pm		6h pm		8h pm		10h pm		Midnight		Mean
3062m.																		
	mps		mps		mps		mps		mps		mps		mps		mps		mps	mps
..	..	WSW	2,2	..	..	WSW	4,2	..	..	WSW	6,4	WSW	11,2	SW	8,4	SW	11,8	9,50
..	..	W	16,5	..	..	W	16,9	..	..	W	21,8	W	15,8	NNW	14,1	NW	16,1	15,44
..	..	W	9,2	..	..	W	12,1	..	..	SW	10,8	SW	11,8	WSW	10,8	WSW	14,7	12,44
..	..	WSW	8,0	..	..	WSW	8,6	..	..	NW	11,1	WNW	7,9	W	8,5	WSW	12,4	10,54
..	..	SW	3,0	..	..	W	5,2	..	..	W	9,1	W	13,1	W	15,2	W	17,0	10,83
..	..	SSW	2,8	..	..	NW	1,6	..	..	WSW	2,0	W	5,7	W	6,8	W	11,3	8,44
..	..	SSW	0,8	..	..	W	0,7	..	..	NE	2,2	N	4,9	NNW	5,2	NW	6,1	6,11
..	..	WSW	2,1	..	..	S	2,8	..	..	SSE	2,4	S	4,2	SW	10,2	WSW	5,9	4,84
..	..	W	3,8	..	..	WNW	4,5	..	..	NW	13,1	WNW	15,4	NW	22,5	NW	18,2	12,08
..	..	N	7,6	..	..	N	4,1	..	..	W	1,7	NW	8,0	WNW	12,2	NW	15,2	12,17
..	..	WNW	2,6	..	..	W	2,2	..	..	W	1,7	W	5,4	NW	12,8	NW	23,8	7,82
..	..	WSW	2,2	..	..	W	2,2	..	..	WNW	0,8	NNW	3,0	W	7,6	W	3,0	5,70
	..		7,65		..		7,82		..		9,95		11,71		13,10		14,89	11,70
..	..	NNW	3,7	..	..	NW	4,1	..	..	WNW	10,6	WNW	12,2	W	12,2	W	10,8	9,33
..	..	N	2,2	..	..	W	4,5	..	..	NNW	6,5	NW	11,8	NW	13,8	NW	9,7	9,01
..	..	WNW	2,4	..	..	WSW	1,4	..	..	W	0,5	SSW	2,7	W	4,2	WNW	7,0	5,62
..	..	N	4,2	..	..	W	6,0	..	..	NNW	4,3	WSW	7,8	WSW	12,4	WSW	14,0	6,85
..	..	SW	1,6	..	..	SW	2,0	..	..	NNW	6,9	NW	10,0	NW	8,7	W	10,1	11,58
..	..	W	2,5	..	..	ESE	1,2	..	..	SSW	1,8	W	6,8	W	5,2	W	8,1	4,51
..	..	WSW	5,2	..	..	W	13,5	..	..	W	12,6	W	18,0	W	22,3	SW	9,7	10,94
..	..	SW	7,9	..	..	SW	10,6	..	..	SSW	15,5	SSW	23,4	SSW	14,6	SW	19,5	12,47
..	..	W	12,1	..	..	WSW	14,7	..	..	WSW	9,4	SW	6,2	SSW	12,8	SSW	15,0	12,88
..	..	SW	15,2	..	..	SW	15,2	..	..	SW	14,0	SW	13,1	WSW	12,4	SW	14,7	18,81
..	..	SW	11,9	..	..	SSW	15,6	..	..	SSW	16,0	SSW	19,4	SSW	18,2	SSW	17,7	14,30
..	..	S	6,3	..	..	S	10,8	..	..	S	9,7	SSE	7,6	SE	9,2	SE	11,2	11,42
	..		7,27		..		7,95		..		9,68		11,67		12,84		14,17	11,41
DAKE 1200m.																		
..	..	W	11,7	..	..	WNW	12,7	..	..	WNW	4,6	NW	7,0	NW	6,8	NW	7,3	10,1
..	..	NW	8,4	..	..	NW	9,8	..	..	NW	15,5	NNW	14,0	NW	13,5	NW	13,0	10,8
..	..	S	1,8	..	..	SSE	1,7	..	..	SSE	4,3	SE	5,9	SE	7,9	SE	8,4	6,3
..	..	SW	2,0	..	..	WSW	2,5	..	..	W	2,2	WSW	3,3	WSW	8,6	SW	3,7	4,5
..	..	NW	1,4	..	..	W	1,6	..	..	SW	1,3	ESE	2,2	SW	2,3	NW	12,3	3,1
..	..	NE	3,3	..	..	SE	9,2	..	..	SE	5,1	SE	5,7	SE	8,6	E	6,5	4,6
..	..	SE	4,7	..	..	SE	3,6	..	..	E	4,9	ENE	6,0	E	8,4	E	8,4	6,9
..	..	SSE	21,0	..	..	SSE	18,5	..	..	SSE	21,2	SSE	20,7	SSE	20,3	SSE	22,3	17,8
..	..	NW	7,1	..	..	NW	8,6	..	..	NNW	10,5	NNW	12,8	NNW	10,0	NNW	10,7	12,4
..	..	NW	13,1	..	..	NW	6,9	..	..	NW	6,4	NW	15,3	NW	12,8	NW	13,8	8,2
..	..	NW	4,9	..	..	NW	9,8	..	..	NNW	8,7	NW	8,4	NNW	2,7	NNW	3,5	7,4
..	..	SE	5,0	..	..	SSE	3,5	..	..	SE	5,7	SE	6,0	SE	7,8	ESE	11,4	4,6
..	..	NW	0,6	..	..	NW	3,2	..	..	NNW	5,8	NNW	6,8	NNW	5,3	NNW	9,0	6,0
..	..	NW	12,5	..	..	NNW	7,9	..	..	NW	10,9	NW	14,2	NW	16,1	NNW	14,7	11,4
..	..	NNW	6,8	..	..	NW	12,1	..	..	NW	12,0	NNW	10,2	NNW	7,3	NE	2,0	9,0
..	..	SE	7,0	..	..	SE	8,3	..	..	SE	10,3	SE	15,1	SE	14,6	SE	15,0	8,6
..	..	SSE	8,2	..	..	S	8,7	..	..	S	11,6	S	8,6	SSE	9,3	SSW	8,0	11,6
..	..	NNW	1,8	..	..	NNW	1,2	..	..	N	3,2	NNE	2,1	NNE	1,7	N	4,5	2,7
..	..	N	2,3	..	..	N	4,9	..	..	N	4,2	N	7,6	N	5,9	N	9,2	4,3
..	..	NW	3,2	..	..	WNW	3,0	..	..	NW	5,3	NW	2,1	NNW	4,5	NW	7,4	7,3
..	..	NW	12,1	..	..	NW	6,9	..	..	NW	13,4	NW	11,4	NNW	7,6	NNW	10,7	9,8
..	..	SSE	9,3	..	..	SE	8,4	..	..	SE	6,4	SSE	8,9	SSE	1,8	SSE	7,1	6,0
..	..	SE	2,9	..	..	SSE	1,9	..	..	SE	1,6	SE	1,6	SE	4,0	SE	1,1	3,0
..	..	NW	1,7	..	..	WNW	3,9	..	..	WNW	5,2	NW	6,1	NNW	6,9	NNW	6,9	3,4
..	..	WNW	3,2	..	..	NW	4,4	..	..	NW	9,4	NW	11,5	NW	10,6	WNW	13,7	7,4
..	..	NW	4,0	..	..	NW	7,1	..	..	NNW	7,8	NNW	9,9	NNW	10,9	NNW	11,3	7,4
..	..	NW	3,0	..	..	NW	5,0	..	..	NW	5,9	NW	4,8	NW	3,2	—	..	5,1
..	..	NW	5,1	..	..	NW	9,2	..	..	NW	11,2	NW	14,5	NNW	12,1	NW	5,9	8,2
..	..	NW	16,2	..	..	NW	18,8	..	..	NW	18,9	NW	14,7	NNW	14,5	NNW	13,9	15,6
..	..	NNW	6,0	..	..	NNW	7,0	..	..	NW	11,8	N	5,4	N	7,4	NNW	12,5	7,4
	..		6,4		..		7,0		..		8,2		8,8		8,3		8,7	7,7

Month & Year	Day	2h am		4h am		6h am		7h am		8h am		9h am		10h am		11h am		Noon		1h pm	
																					YAMANAKA
			mps		mps		mps		mps		mps		mps		mps		mps		mps		mps
VIII 1889	1	...	...	...	...	—	0.0	—	0.0	—	0.0	—	0.0	—	0.0	—	0.0	SE	2.4	SE	3.0
	2	...	...	...	...	—	0.0	—	0.0	—	0.0	SE	0.9	..	1.5	SE	2.5	SE	3.0	SE	3.7
	3	—	0.0	...	...	—	0.0	—	0.0	—	0.0	—	0.0	SSE	0.6	SE	1.6	SE	1.9	SE	2.4
	4	—	0.0	...	...	—	0.0	—	0.0	—	0.0	SSE	1.1	SE	1.6	E	3.6	SSE	3.4	SE	4.2
	5	—	0.0	...	...	—	0.0	E	2.2	E	2.9	SE	2.1	SE	2.5	SE	2.2	SE	2.7	SSE	0.7
	6	—	0.0	...	...	—	0.0	—	0.0	—	0.0	SW	0.7	W	0.8	W	1.2	SSW	2.4	SE	2.1
	7	—	0.0	...	...	—	0.0	—	0.0	NE	1.4	SE	2.7	E	3.7	ESE	3.7	SE	4.1	SE	3.2
	8	...	...	...	...	—	0.0	—	0.0	—	0.0	—	0.0	SE	2.9	SE	2.2	SSE	1.2	S	3.7
	9	SE	1.0	...	...	—	0.0	—	0.0	SSE	0.6	SE	0.8	SSE	2.6	SE	3.4	SE	4.0	SW	4.0
	10	—	0.0	...	...	—	0.0	—	0.0	—	0.0	SE	1.2	SE	2.4	S	2.3	SSE	3.0	SSE	3.2
	11	—	0.0	...	...	—	0.0	—	0.0	E	0.7	SE	2.3	—	0.0	SE	3.4	SSE	3.7	SSE	3.3
	12	—	0.0	...	...	—	0.0	—	0.0	—	0.0	—	0.0	S	0.9	SE	2.3	SE	2.5	SE	4.3
	13	—	0.0	...	...	—	0.0	—	0.0	—	0.0	E	2.0	E	2.3	SE	3.0	SSW	3.4	E	4.0
	14	—	0.0	...	...	—	0.0	S	0.6	E	1.1	—	0.0	SE	6.8	SE	1.9	SE	6.2	SE	1.4
	15	—	0.0	...	...	—	0.0	E	1.2	E	1.3	SE	2.7	SE	3.2	ESE	2.7	SE	3.8	SE	8.3
	16	—	0.0	...	...	—	0.0	—	0.0	E	1.5	ESE	2.0	SE	3.3	SE	3.2	SE	3.5	SE	4.2
	17	—	0.0	...	...	—	0.0	—	0.0	E	0.5	E	3.2	E	3.0	SE	1.6	E	2.5	E	3.4
	18	—	0.0	...	...	—	0.0	..	..	SE	2.3	SE	1.2	E	1.4	E	1.4	SE	2.1	SE	0.9
	19	SE	6.1	...	...	SE	3.7	SE	4.8	ESE	6.0	ESE	7.2	ESE	7.9	ESE	4.2	SSE	8.9	SSE	3.2
	20	SE	6.7	...	...	SSE	5.5	SSE	5.9	SSE	7.1	SE	6.3	SSE	5.9	SSE	6.9	S	7.9	S	7.7
	21	S	10.6	...	...	SW	6.9	S	2.1	E	2.6	ESE	4.0	E	4.0	E	2.7	E	4.0	E	3.8
	22	—	0.0	...	...	—	0.0	—	0.0	—	0.0	SE	0.9	E	2.1	E	2.4	E	1.2	E	3.0
	23	—	0.0	...	...	—	0.0	—	0.0	—	0.0	—	0.0	—	0.0	SE	1.5	SE	2.1	S	2.1
	24	—	0	...	...	—	0	—	0	—	0	—	0	—	0	SE	1	SE	1	SE	1
	25	—	0	...	...	—	0	—	0	—	0	—	0	S	1	SE	2	SE	2	SE	1
	26	—	0	...	...	—	0	SE	1	—	0	SE	1	SSE	2	SE	2	SE	3	SE	3
	27	—	0	...	...	—	0	—	0	—	0	—	0	E	1	—	0	—	0	E	1
	28	—	0	...	...	—	0	—	0	—	0	—	0	—	0	—	0	—	0	—	0
	29	—	0	...	...	—	0	—	0	—	0	—	0	—	0	—	0	—	0	—	0
	30	—	0	...	...	—	0	—	0	—	0	—	0	—	0	—	0	—	0	—	0
	31	—	0	...	...	—	0	—	0	—	0	—	0	—	0	—	0	SE	1	—	0
IX 1889	1	—	0	...	...	—	0	...	...	...	...	...	...	E	1	...	...	...	...	...	...
	2	S	1	...	...	SSE	2	...	...	...	...	...	...	S	1	...	...	...	...	...	...
	3	—	0	...	...	—	0	...	...	...	...	...	...	—	0	...	...	...	...	...	...
	4	—	0	...	...	N	1	...	...	...	...	...	...	N	1	...	...	...	...	...	...
	5	—	0	...	...	—	0	...	...	...	...	...	...	S	1	...	...	...	...	...	...
	6	E	3	...	...	E	2	...	...	...	...	...	...	E	2	...	...	...	...	...	...
	7	—	0	...	...	E	1	...	...	...	...	...	...	SE	1	...	...	...	...	...	...
																					KUROSAWA
VIII 1891	1	NW	1.0	—	0.2	—	0.1	...	...	—	0.0	...	...	NW	2.5	...	...	SE	4.6	...	...
	2	—	0.0	—	0.2	—	0.3	...	...	NNW	1.2	...	...	S	2.7	...	...	SSE	1.9	...	...
	3	—	0.0	—	0.2	—	0.0	...	...	—	0.0	...	...	SE	1.9	...	...	NE	1.0	...	...
	4	S	1.2	SSE	1.2	NNE	0.5	...	...	S	2.0	...	...	SE	2.2	...	...	SW	1.0	...	...
	5	—	0.1	—	0.4	—	0.0	...	...	—	0.0	...	...	—	0.2	...	...	S	3.6	...	...
	6	—	0.0	—	0.0	—	0.0	...	...	—	0.0	...	...	W	1.3	...	...	S	1.4	...	...
	7	—	0.0	—	0.0	—	0.0	...	...	SW	1.7	...	...	SW	2.3	...	...	—	0.1	...	...
	8	—	0.2	—	0.0	—	0.0	...	...	—	0.1	...	...	NE	2.6	...	...	NW	0.9	...	...
	9	—	0.0	—	0.1	—	0.0	...	...	—	0.0	...	...	—	0.0	...	...	N	0.7	...	...
	10	—	0.0	—	0.2	—	0.1	...	...	—	0.1	...	...	SSE	1.7	...	...	SSE	3.3	...	...
	11	—	0.3	N	1.0	—	0.0	...	...	W	1.2	...	...	N	1.0	...	...	SE	1.7	...	...
	12	—	0.0	—	0.0	—	0.0	...	...	NNW	0.5	...	...	SW	2.7	...	...	S	2.5	...	...
	13	—	0.4	—	0.0	NW	1.2	...	...	—	0.2	...	...	WSW	0.7	...	...	W	1.2	...	...
	14	NW	1.2	—	0.0	—	0.0	...	...	—	0.0	...	...	S	3.2	...	...	SSW	1.8	...	...
	15	—	0.0	—	0.0	NNW	0.7	...	...	—	0.2	...	...	S	1.2	...	...	SE	5.7	...	...
	16	—	0.1	—	0.0	—	0.0	...	...	—	0.3	...	...	SSW	4.2	...	...	S	4.2	...	...
	17	S	2.6	E	7.4	SE	6.9	...	...	SE	8.0	...	...	E	2.0	...	...	SW	1.3	...	...
	18	—	0.0	—	0.0	—	0.0	...	...	ESE	1.0	...	...	—	0.4	...	...	SE	2.4	...	...
	19	—	0.0	—	0.0	—	0.0	...	...	—	0.0	...	...	—	0.0	...	...	—	0.0	...	...

2h pm		3h pm		4h pm		5h pm		6h pm		7h pm		8h pm		9h pm		10h pm		11h pm		Midnight		Mean
990m.																						
	mps		mps		mps		mps		mps		mps		mps		mps		mps		mps		mps	mps
SE	2.1	SE	2.4	SE	1.8	SE	1.9	SE	2.1	..	1.4	—	0.0	—	0.0	—	0.0	..	..	..	..	0.70
SE	3.4	SE	3.9	SE	2.4	SE	3.3	SE	2.0	SE	1.0	—	0.0	—	0.0	—	0.0	..	..	..	..	1.15
SE	1.9	SSE	2.4	SSE	2.5	SE	2.1	SE	2.8	—	0.0	—	0.0	—	0.0	—	0.0	..	..	..	..	0.88
SE	2.9	SE	3.5	SE	2.6	S	2.1	S	1.5	S	1.0	S	0.6	—	0.0	—	0.0	..	..	..	..	1.00
S	3.3	E	2.5	SSE	3.1	SSE	2.1	S	1.0	S	0.5	—	0.0	—	0.0	—	0.0	..	..	..	..	1.13
SSE	3.3	SE	3.1	SSE	2.6	SE	8.8	SE	2.3	S	1.4	S	0.7	—	0.0	—	0.0	..	..	..	..	1.07
S	5.0	S	5.5	S	5.5	S	4.6	S	2.9	S	2.7	S	0.8	—	0.0	—	0.0	..	..	..	..	1.93
S	2.1	S	2.7	S	4.0	SE	4.7	SE	2.4	—	0.0	SE	1.6	—	0.0	—	0.0	..	..	..	..	1.23
ESE	4.3	SE	3.0	SSE	3.5	S	1.7	S	1.7	E	0.5	—	0.0	—	0.0	—	0.0	—	0.0	—	0.0	1.60
SE	2.8	SSE	2.1	SSE	3.0	E	3.0	SE	1.9	S	1.8	S	1.3	—	0.0	—	0.0	—	0.0	—	0.0	1.48
SSE	3.3	SE	1.0	SSE	5.5	SSE	3.6	SE	1.6	SSE	1.4	SSE	0.8	SSE	0.9	SSE	0.9	—	0.0	—	0.0	0.97
SE	3.9	SE	3.2	SSE	2.8	SSE	1.7	SSW	1.6	S	0.7	—	0.0	—	0.0	—	0.0	—	0.0	—	0.0	0.92
S	4.5	E	3.2	ESE	2.7	S	2.5	S	1.3	—	0.0	—	0.0	—	0.0	—	0.0	—	0.0	—	0.0	1.35
SE	1.5	—	0.0	SE	7.2	SE	2.3	S	1.3	S	0.8	—	0.0	—	0.0	—	0.0	—	0.0	—	0.0	1.00
—	0.0	..	2.5	SE	3.2	S	2.4	SE	2.5	E	1.9	SE	1.4	—	0.0	—	0.0	—	0.0	—	0.0	0.95
SE	2.2	SE	2.4	SE	3.5	E	2.5	SE	2.8	E	1.4	—	0.0	—	0.0	—	0.0	..	..	—	0.0	1.38
SE	4.4	SE	3.0	SE	3.8	E	3.6	E	3.0	E	4.7	E	2.3	—	..	—	0.0	..	..	..	..	1.73
—	0.0	—	0.0	SE	0.6	SE	1.2	SE	3.6	SE	5.3	SE	4.8	SE	4.0	SE	4.9	SE	8.5	SE	5.4	1.65
ESE	5.1	SE	4.5	SSE	4.8	SE	2.7	SE		SE	..	SE	1.4	SE	5.5	SE	6.5	..	..	SE	2.8	4.88
S	11.6	S	6.0	S	8.3	S	17.4	S	5.3	S	9.1	S	9.1	S	7.8	S	8.9	S	11.3	S	8.9	7.32
E	4.0	SW	3.5	WSW	4.1	S	3.4	E	2.9	E	1.1	—	0.0	—	0.0	—	0.0	..	..	—	0.0	4.73
E	3.2	E	4.0	S	3.0	SE	2.0	..	1.0	—	0.0	—	0.0	—	0.0	—	0.0	..	..	—	0.0	1.05
S	3.0	S	3.7	S	2.7	S	0.6	—	0.0	—	0.0	—	0.0	—	0.0	—	0.0	—	0.0	—	0.0	0.25
SE	**1**	SE	**1**	—	**0**	—	**0**	—	**0**	—	**0**	—	**0**	—	**0**	—	**0**	—	**0**	—	**0**	..
SE	**2**	SE	**2**	SE	**1**	SE	**1**	SE	**1**	SE	**1**	—	**0**	—	**0**	—	**0**	—	**0**	—	**0**	..
SE	**3**	SE	**2**	—	**0**	—	**0**	—	**0**	—	**0**	—	**0**	—	**0**	—	**0**	—	**0**	SE	**1**	..
—	**0**	—	**0**	—	**0**	—	**0**	—	**0**	—	**0**	—	**0**	—	**0**	—	**0**	—	**0**	—	**0**	..
—	**0**	—	**0**	—	**0**	—	**0**	—	**0**	—	**0**	—	**0**	—	**0**	—	**0**	—	**0**	—	**0**	..
—	**0**	—	**0**	—	**0**	—	**0**	—	**0**	—	**1**	—	**0**	—	**0**	—	**0**	—	**0**	—	**0**	..
—	**0**	—	**0**	—	**0**	—	**0**	—	**0**	—	**1**	—	**0**	—	**0**	—	**0**	—	**0**	—	**0**	..
SSE	**1**	S	**1**	S	**1**	—	**0**	—	**0**	—	**1**	—	**0**	—	**0**	S	**1**	—	**0**	—	**1**	..
ESE	**2**	..	..	..	..	..	..	S	**1**	..	..	..	..	..	..	—	**0**	..	..	..	..	..
E	**1**	..	..	..	..	..	..	N	**1**	..	..	..	..	..	..	—	**0**	..	..	..	..	..
—	**0**	..	..	..	..	..	..	—	**0**	..	..	..	..	..	..	—	**0**	..	..	..	..	..
E	**1**	..	..	..	..	..	..	SSW	**1**	..	..	..	..	..	..	—	**0**	..	..	..	..	..
S	**1**	..	..	..	..	..	..	E	**1**	..	..	..	..	..	..	—	**0**	..	..	..	..	..
S	**1**	..	..	..	..	..	..	E	**1**	..	..	..	..	..	..	—	**0**	..	..	..	..	..
ESE	**1**	..	..	..	..	..	..	E	**1**	..	..	..	..	..	..	—	**0**	..	..	..	..	..
832m.																Black figures refer to estimated forces.						
S	3.9	..	..	S	3.6	..	..	SE	2.0	..	..	—	0.3	..	..	—	0.1	..	..	—	0.0	1.52
SE	1.7	..	..	SE	2.4	..	..	—	0.0	..	..	—	0.0	..	..	—	0.0	..	..	—	0.0	0.87
SSE	4.2	..	..	SE	5.2	..	..	SE	1.5	..	..	SE	0.6	..	..	ESE	2.6	..	..	SE	2.0	1.60
SE	1.0	..	..	—	0.4	..	..	SSW	0.5	..	..	—	0.0	..	..	N	0.8	..	..	WNW	0.5	0.94
—	0.0	..	..	—	0.3	..	..	—	0.0	..	..	—	0.0	..	..	—	0.0	..	..	—	0.0	0.38
E	3.0	..	..	S	2.2	..	..	—	0.4	..	..	—	0.2	..	..	—	0.0	..	..	SSW	0.5	0.75
E	1.3	..	..	ENE	0.8	..	..	—	0.2	..	..	—	0.1	..	..	NNE	0.6	..	..	—	0.0	0.59
S	2.5	..	..	NW	1.2	..	..	SE	2.2	..	..	—	0.0	..	..	—	0.0	..	..	—	0.2	0.82
—	0.2	..	..	S	2.5	..	..	SSE	1.5	..	..	N	0.6	..	..	N	0.6	..	..	—	0.0	0.52
S	3.7	..	..	WNW	2.0	..	..	SW	1.2	..	..	NE	1.0	..	..	—	0.0	..	..	—	0.0	1.11
WNW	1.7	..	..	SSW	1.4	..	..	—	0.4	..	..	—	0.0	..	..	—	0.1	..	..	—	0.2	0.77
S	3.6	..	..	SSE	2.9	..	..	SSE	1.2	..	..	—	0.0	..	..	ESE	0.8	..	..	NW	0.6	1.23
NW	0.8	..	..	S	4.2	..	..	E	0.7	..	..	—	0.0	..	..	W	0.6	..	..	NW	1.5	0.94
ESE	3.2	..	..	SE	1.6	..	..	SE	1.9	..	..	—	0.2	..	..	—	0.0	..	..	—	0.0	1.09
SW	1.4	..	..	WSW	1.3	..	..	SW	0.7	..	..	—	0.0	..	..	—	0.2	..	..	—	0.1	0.97
SSE	5.8	..	..	SE	1.8	..	..	SW	2.5	..	..	N	2.2	..	..	N	1.5	..	..	ENE	5.1	2.31
SSE	2.3	..	..	SE	1.3	..	..	—	0.0	..	..	—	0.1	..	..	—	0.0	..	..	—	0.0	2.66
—	0.0	..	..	S	2.3	..	..	—	0.0	..	..	—	0.0	..	..	—	0.4	..	..	—	0.2	0.56
ESE	1.2	..	..	SW	2.0	..	..	E	5.7	..	..	—	0.1	..	..	—	0.1	..	..	—	0.0	0.76

Month & Year	Day	2h am		4h am		6h am		7h am		8h am		9h am		10h am		11h am		Noon		1h pm	
KUROSAWA																					
			mps		mps		mps		mps		mps		mps		mps		mps		mps		mps
VIII 1891	20	—	0,0	—	0,0	—	0,1	..	..	—	0,0	..	..	—	0,2	..	..	NW	1,2	..	..
	21	—	0,0	—	0,0	—	0'0	..	..	E	0,5	..	..	E	1,2	..	..	—	0,4	..	..
	22	—	0,0	—	0,3	—	0,0	..	..	—	0,3	..	..	—	0,2	..	..	ENE	1,0	..	..
	23	—	0,0	—	0,0	—	0,0	..	..	—	0,2	..	..	—	0,0	..	..	E	1,4	..	..
	24	—	0,0	—	0,0	—	0,0	..	..	—	0,0	..	..	E	1,4	..	..	S	2,2	..	..
	25	—	0,0	ENE	0,9	E	1,4	..	..	—	0,1	..	..	E	1,0	..	..	WNW	2,0	..	..
	26	—	0,0	—	0,1	NE	0,5	..	..	—	0,2	..	..	S	0,5	..	..	NE	1,7	..	..
	27	—	0,0	—	0,0	—	0,0	..	..	—	0,0	..	..	S	1,1	..	..	SSW	3,7	..	..
	28	—	0,0	—	0,4	—	0,0	..	..	—	0,0	..	..	S	1,5	..	..	S	1,6	..	..
	29	—	0,2	—	0,2	—	0,2	..	..	—	0,0	..	..	S	1,7	..	..	SSW	0,7	..	..
	30	—	0,0	—	0,0	—	0,0	..	..	NE	0,7	..	..	—	0,1	..	..	SSW	2,1	..	..
	31	—	0,0	—	0,0	—	0,0	..	..	—	0,0	..	..	SE	1,9	..	..	S	1,9	..	..
	Mean		0,24		0,41		0,39	..	..		0,60	..	..		1,41	..	..		1,91	..	..
IX 1891	1	—	0,0	—	0,4	—	0,2	..	..	—	0,0	..	..	S	0,6	..	..	ESE	2,8	..	..
	2	—	0,0	—	0,2	N	0,9	..	..	—	0,0	..	..	—	0,0	..	..	SSW	0,9	..	..
	3	—	0,2	—	0,0	—	0,0	..	..	—	0,0	..	..	S	1,3	..	..	S	1,6	..	..
	4	N	0,9	—	0,0	N	1,1	..	..	—	0,0	..	..	E	5,1	..	..	S	1,5	..	..
	5	—	0,0	—	0,1	—	0,1	..	..	—	0,1	..	..	E	2,1	..	..	S	3,9	..	..
	6	—	0,0	N	2,2	N	0,5	..	..	—	0,0	..	..	S	2,5	..	..	S	2,6	..	..
	7	—	0,1	—	0,0	—	0,0	..	..	—	0,0	..	..	S	1,3	..	..	SE	1,4	..	..
	8	—	0,0	—	0,0	—	0,1	..	..	—	0,0	..	..	S	1,8	..	..	E	1,2	..	..
	9	NE	1,8	—	0,0	—	0,0	..	..	SE	0,8	..	..	ESE	1,6	..	..	SE	2,7	..	..
	10	—	0,0	—	0,0	—	0,0	..	..	—	0,0	..	..	SE	5,1	..	..	SE	4,4	..	..
	11	—	0,0	—	0,0	—	0,0	..	..	SE	1,5	..	..	SE	2,2	..	..	SE	5,6	..	..
	12	—	0,0	—	0,0	—	0,0	..	..	—	0,0	..	..	SE	2,2	..	..	SE	1,2	..	..
General Mean			0,24		0,36		0,35	..	..		0,49	..	..		1,62	..	..		2,07	..	..
YOKKAICHI																					
IX 1888	4	NNW	0,7	..	..	SW	4,7	..	..	..	..	..	..	W	5,7	..	..	..	..	..	..
	5	NW	1,1	..	..	NW	0,9	..	..	..	..	..	..	NNW	2,0	..	..	..	..	..	..
	6	NW	2,6	..	..	NW	1,9	..	..	..	..	..	..	SE	3,3	..	..	..	..	..	..
	7	NW	1,1	..	..	NW	0,7	..	..	..	..	..	..	E	2,4	..	..	..	..	..	..
	8	NW	0,8	..	..	NW	0,8	..	..	..	..	..	..	WNW	0,9	..	..	..	..	..	..
	9	WNW	2,5	..	..	NW	1,6	..	..	..	..	..	..	N	2,7	..	..	..	..	..	..
	10	WNW	1,2	..	..	NNE	2,0	..	..	..	..	..	..	NW	1,7	..	..	..	..	..	..
	11	WNW	3,5	..	..	W	0,6	..	..	..	..	..	..	NNW	2,7	..	..	..	..	..	..
	12	SE	19,4	..	..	S	5,0	..	..	..	..	..	..	SW	4,9	..	..	..	..	..	..
	13	NW	4,1	..	..	NW	2,3	..	..	..	..	..	..	WSW	1,4	..	..	..	..	..	..
	14	NW	4,3	..	..	NW	1,1	..	..	..	..	..	..	ENE	1,9	..	..	..	..	..	..
	15	WNW	2,3	..	..	NW	0,8	..	..	..	..	..	..	ENE	2,0	..	..	..	..	..	..
	16	S	3,9	..	..	—	0,0	..	..	..	..	..	..	—	0,3	..	..	..	..	..	..
	17	—	0,4	..	..	NW	2,0	..	..	..	..	..	..	NW	5,0	..	..	..	..	..	..
	18	—	0,3	..	..	W	0,8	..	..	..	..	..	..	NE	3,4	..	..	..	..	..	..
	19	NW	1,4	..	..	NW	2,1	..	..	..	..	..	..	NW	2,7	..	..	..	..	..	..
	20	SE	12,1	..	..	SSE	8,4	..	..	..	..	..	..	SE	11,6	..	..	..	..	..	..
	21	W	4,7	..	..	NW	1,1	..	..	..	..	..	..	NW	0,5	..	..	..	..	..	..
	22	S	0,5	..	..	NW	2,9	..	..	..	..	..	..	N	1,6	..	..	..	..	..	..
	23	NW	3,3	..	..	W	0,7	..	..	..	..	..	..	NW	5,4	..	..	..	..	..	..
	24	W	0,8	..	..	W	1,0	..	..	..	..	..	..	NW	1,3	..	..	..	..	..	..
	25	WNW	0,7	..	..	W	0,8	..	..	..	..	..	..	NW	0,7	..	..	..	..	..	..
	26	NW	0,6	..	..	NW	1,9	..	..	..	..	..	..	ENE	2,9	..	..	..	..	..	..
	27	W	1,8	..	..	WSW	0,9	..	..	..	..	..	..	SE	1,4	..	..	..	..	..	..
	28	W	2,0	..	..	NW	0,7	..	..	..	..	..	..	S	1,7	..	..	..	..	..	..
	29	NW	2,8	..	..	NW	0,7	..	..	..	..	..	..	ENE	3,9	..	..	..	..	..	..
	30	NW	2,6	..	..	NW	2,7	..	..	..	..	..	..	E	4,9	..	..	..	..	..	..
X 1888	1	NNW	2,0	..	..	WNW	0,7	..	..	..	..	..	..	NW	1,1	..	..	..	..	..	..
	2	NNW	9,2	..	..	NW	5,7	..	..	..	..	..	..	N	7,5	..	..	..	..	..	..
	3	NW	0,4	..	..	NW	1,0	..	..	..	..	..	..	N	3,9	..	..	..	..	..	..
	Mean		3,1	..	..		1,9	..	..	..	..	..	..		3,0	..	..	..	..	..	..

2h pm		3h pm		4h pm		5h pm		6h pm		7h pm		8h pm		9h pm		10h pm		11h pm		Midnight		Mean
832m.																						
	mps		mps		mps		mps		mps		mps		mps		mps		mps		mps		mps	mps
—	0,0	..	..	E	3,7	..	..	—	0,0	..	..	—	0,0	..	..	—	0,0	..	..	E	1,4	0,55
E	0,6	..	..	E	1,2	..	..	—	0,4	..	..	E	1,7	..	..	SE	0,6	..	..	SE	1,2	0,65
E	1,8	..	..	ESE	2,4	..	..	—	0,2	..	..	—	0,0	..	..	—	0,9	..	..	—	0,0	0,52
—	0,9	..	..	—	0,0	..	..	—	0,0	..	..	—	0,0	..	..	E	0,8	..	..	—	0,0	0,20
NE	3,2	..	..	E	1,5	..	..	—	0,0	..	..	—	0,0	..	..	—	0,0	..	..	NE	0,7	0,75
NW	1,3	..	..	NNE	0,7	..	..	NNE	0,7	..	..	NNE	0,7	..	..	—	0,3	..	..	—	0,0	0,76
S	1,8	..	..	SW	1,2	..	..	—	0,0	..	..	—	0,2	..	..	—	0,1	..	..	—	0,0	0,53
SSW	3,1	..	..	S	2,6	..	..	E	1,6	..	..	—	0,4	..	..	—	0,0	..	..	—	0,0	1,04
S	4,1	..	..	WSW	1,6	..	..	E	1,7	..	..	—	0,0	..	..	—	0,0	..	..	—	0,4	0,94
SW	2,2	..	..	NW	2,6	..	..	N	1,2	..	..	W	0,7	..	..	—	0,0	..	..	—	0,0	0,81
S	2,2	..	..	SW	0,6	..	..	NE	1,3	..	..	—	0,2	..	..	—	0,0	..	..	—	0,0	0,62
S	1,8	..	..	SE	3,2	..	..	SE	0,7	..	..	—	0,0	..	..	—	0,0	..	..	N	1,2	0,89
	2,05		..		1,96		..		0,98		..		0,20		..		0,31		..		0,50	0,92
S	3,2	..	..	S	3,1	..	..	E	1,7	..	..	—	0,0	..	..	—	0,0	..	..	ENE	0,5	1,04
—	0,2	..	..	S	3,4	..	..	—	0,0	..	..	—	0,0	..	..	—	0,0	..	..	—	0,0	0,47
SSE	0,7	..	..	W	1,7	..	..	NE	0,7	..	..	—	0,0	..	..	—	0,0	..	..	NNW	1,0	0,60
SE	4,6	..	..	SSE	4,5	..	..	N	2,2	..	..	—	0,3	..	..	—	0,0	..	..	—	0,0	1,68
S	3,0	..	..	SE	2,1	..	..	—	0,0	..	..	—	0,0	..	..	—	0,0	..	..	—	0,0	0,95
SE	3,8	..	..	SE	4,2	..	..	SSE	1,3	..	..	—	0,0	..	..	—	0,2	..	..	—	0,0	1,44
SE	2,7	..	..	ESE	1,9	..	..	E	1,5	..	..	—	0,2	..	..	—	0,2	..	..	—	0,0	0,77
—	0,0	..	..	—	0,2	..	..	—	0,0	..	..	—	0,2	..	..	—	0,4	..	..	—	0,0	0,32
SE	2,7	..	..	SE	3,0	..	..	—	0,0	..	..	NE	0,8	..	..	—	0,0	..	..	—	0,0	1,12
S	6,0	..	..	SE	4,7	..	..	—	0,0	..	..	—	0,0	..	..	—	0,0	..	..	—	0,0	1,68
S	2,4	..	..	S	2,2	..	..	NE	0,5	..	..	NE	1,6	..	..	—	0,0	..	..	—	0,0	1,33
SE	2,8	..	..	S	1,5	..	..	—	0,3	..	..	—	0,0	..	..	—	0,0	..	..	—	0,0	0,67
	2,28		..		2,17		..		0,90		..		0,29		..		0,26		..		0,40	0,95
4m.																						
S	5,9	..	..	..	..	..	..	SSW	1,7	..	..	..	..	..	..	WNW	2,1	..	..	..	..	3,2
NW	6,4	..	..	..	..	..	..	NNW	6,9	..	..	..	..	..	..	NW	4,6	..	..	..	..	3,7
SE	2,1	..	..	..	..	..	..	SSE	4,8	..	..	..	..	..	..	NW	4,2	..	..	..	..	2,9
S	2,3	..	..	..	..	..	..	W	1,0	..	..	..	..	..	..	NW	2,8	..	..	..	..	1,9
W	1,1	..	..	..	..	..	..	NW	2,7	..	..	..	..	..	..	NNW	1,2	..	..	..	..	1,5
NNW	3,1	..	..	..	..	..	..	WSW	1,2	..	..	..	..	..	..	W	2,1	..	..	..	..	2,1
N	3,6	..	..	..	..	..	..	NNW	3,0	..	..	..	..	..	..	N	6,9	..	..	..	..	3,1
E	2,7	..	..	..	..	..	..	SE	13,7	..	..	..	..	..	..	SE	17,3	..	..	..	..	6,9
SE	2,0	..	..	..	..	..	..	NW	3,4	..	..	..	..	..	..	NW	1,3	..	..	..	..	7,2
NW	11,8	..	..	..	..	..	..	NW	4,4	..	..	..	..	..	..	NNW	3,0	..	..	..	..	3,6
NW	5,0	..	..	..	..	..	..	NW	2,8	..	..	..	..	..	..	SW	1,3	..	..	..	..	3,2
NE	1,3	..	..	..	..	..	..	—	0,2	..	..	..	..	..	..	ENE	1,6	..	..	..	..	1,3
SE	1,1	..	..	..	..	..	..	W	0,8	..	..	..	..	..	..	WNW	2,6	..	..	..	..	1,8
NW	8,0	..	..	..	..	..	..	WNW	5,1	..	..	..	..	..	..	NW	1,3	..	..	..	..	4,3
NNW	9,0	..	..	..	..	..	..	NW	6,9	..	..	..	..	..	..	NW	2,1	..	..	..	..	2,8
—	0,4	..	..	..	..	..	..	SSE	7,7	..	..	..	..	..	..	SE	8,3	..	..	..	..	2,6
SE	8,5	..	..	..	..	..	..	SE	8,3	..	..	..	..	..	..	SE	5,9	..	..	..	..	10,1
N	1,0	..	..	..	..	..	..	W	0,6	..	..	..	..	..	..	W	0,9	..	..	..	..	1,6
N	0,5	..	..	..	..	..	..	NW	1,8	..	..	..	..	..	..	NW	2,0	..	..	..	..	1,7
NW	5,3	..	..	..	..	..	..	NW	3,6	..	..	..	..	..	..	W	0,8	..	..	..	..	3,2
NW	6,3	..	..	..	..	..	..	NW	4,5	..	..	..	..	..	..	NW	4,6	..	..	..	..	3,6
NW	0,7	..	..	..	..	..	..	NE	1,0	..	..	..	..	..	..	—	0,1	..	..	..	..	1,6
S	3,3	..	..	..	..	..	..	SW	1,1	..	..	..	..	..	..	W	1,2	..	..	..	..	1,8
S	3,8	..	..	..	..	..	..	SSW	2,5	..	..	..	..	..	..	W	2,4	..	..	..	..	1,8
NW	6,6	..	..	..	..	..	..	WNW	3,8	..	..	..	..	..	..	NNW	6,6	..	..	..	..	2,9
NW	3,7	..	..	..	..	..	..	NW	2,5	..	..	..	..	..	..	NW	1,0	..	..	..	..	2,8
SE	3,0	..	..	..	..	..	..	SW	0,8	..	..	..	..	..	..	NW	2,3	..	..	..	..	2,3
NW	4,2	..	..	..	..	..	..	NNW	3,4	..	..	..	..	..	..	NW	5,3	..	..	..	..	5,3
NW	8,5	..	..	..	..	..	..	NW	5,4	..	..	..	..	..	..	NW	3,2	..	..	..	..	6,1
NW	4,1	..	..	..	..	..	..	NW	2,2	..	..	..	..	..	..	NW	1,9	..	..	..	..	3,7
	4,2		..		..		..		3,6		..		..		..		3,4		..		..	3,4

Month & Year	Day	2^h am		6^h am		10^h am		2^h pm		6^h pm		10^h pm		Mean
HIGASHI HOBEN 736m.														
VIII 1889			mps		mps		mps		mps		mps		mps	mps
	1	SE	9,0	ESE	9,0	SSE	7,1	SSE	6,4	SSE	4,7	SE	9,2	7,57
	2	SE	6,2	SE	10,1	ESE	4,6	SE	4,7	—	0,1	NNE	8,1	5,62
	3	NE	1,7	N	6,7	N	16,7	N	3,7	N	16,9	NE	15,2	10,15
	4	NE	10,7	NNE	10,6	NNE	10,8	N	14,2	N	6,6	NE	7,4	9,97
	5	N	5,8	N	2,7	SE	3,0	S	5,3	SW	3,9	W	8,5	4,08
	6	SW	7,7	SE	5,2	SSE	6,4	SSE	5,7	SW	2,4	SE	8,0	5,90
	7	SE	4,6	S	5,6	S	6,6	SW	4,2	SSE	3,3	W	2,5	4,47
	8	SE	2,0	SE	5,6	W	1,7	SSE	2,7	SSW	5,5	S	0,6	4,18
	9	SSW	5,5	W	3,9	E	1,7	—	0,4	N	6,0	N	3,4	3,48
	10	N	4,2	SE	1,3	E	1,2	NNE	3,6	N	8,4	N	4,5	3,87
	11	NW	4,7	W	6,8	SE	1,7	NW	4,4	N	10,6	N	3,7	5,32
	12	NNW	10,2	N	8,5	N	5,7	NNE	2,7	N	10,8	N	8,0	7,65
	13	S	6,5	—	0,3	SSE	4,2	SSE	4,0	NE	3,4	SE	6,7	4,25
	14	SE	11,0	SE	12,2	SE	17,1	SE	25,7	SE	16,2	SE	22,1	17,58
	15	SE	20,8	SE	17,4	SE	27,1	SE	22,0	SE	18,8	SE	24,3	21,73
	16	SE	23,7	SE	21,6	SE	19,1	SE	14,9	SE	14,2	SSE	16,3	18,30
	17	SE	13,4	SE	13,4	SE	12,3	SE	10,8	SE	7,5	SE	10,6	11,33
	18	SE	6,9	N	12,2	ENE	5,1	N	10,5	NNE	27,6	N	24,2	14.42
	19	NNE	26,1	N	21,2	N	17,9	N	31,8	N	27,6	N	23,3	24,65
	20	N	19,3	N	20,1	N	18,5	N	13,9	N	12,0	N	9,3	15,52
	21	N	10,5	N	6,5	NNE	2,1	S	1,5	SW	2,5	S	4,2	4,55
	22	ESE	5,2	ESE	8,0	SE	8,3	ESE	4,1	ESE	5,9	SE	8,9	6,73
	23	SE	9,9	SE	9,2	SE	15,6	SE	11,7	ESE	9,0	SE	11,5	11,00
	24	SE	17,6	ESE	18,7	SE	17,8	SE	20,2	SE	22,1	SE	19,7	19,35
	25	SE	21,0	SE	21,4	SE	19,7	SE	11,8	SE	10,2	N	2,1	14,37
	26	N	21,4	NNE	24,7	NNE	27,8	NNE	24,8	N	21,3	N	28,4	25,73
	27	N	25,2	N	23,0	NNE	18,9	NNE	15,5	N	17,7	N	13,1	18,90
	28	N	13,7	NNE	16,8	N	9,6	N	16,1	NNE	13,7	N	9,1	13,17
	29	NNE	13,2	NNW	10,0	N	11,5	N	15,2	NNE	15,4	N	6,7	12,00
	30	N	8,5	N	8,0	N	1,2	N	8,7	N	6,7	ESE	7,1	6,70
	31	SE	9,8	SE	9,2	SE	8,1	SE	8,9	SE	7,2	ESE	11,2	8,77
	Mean		11,79		11,29		10,60		10,62		11,01		10,74	11,01
IX 1889	1	SE	10,1	SE	13,2	SE	11,4	SE	5,7	S	6,4	SSE	14,2	10,17
	2	SE	12,1	SE	12,6	N	6,2	N	11,2	N	12,7	NNE	15,0	11,63
	3	NNE	12,2	NNE	17,3	NE	2,2	NNW	7,5	NNE	10,1	E	2,6	8,65
	4	ESE	6,7	ENE	4,0	SE	6,9	SE	6,7	SE	5,2	SE	10,4	6,65
	5	SE	10,3	SE	6,7	SE	6,5	SSE	6,4	SSE	7,6	SSE	9,5	7,83
	6	SSE	8,3	NNW	7,8	N	7,2	N	13,1	NNW	18,0	NNE	11,4	10,97
	7	N	11,7	NNE	7,7	N	6,3	NNE	7,5	N	13,9	N	14,7	10,30
	8	N	15,2	NNE	13,2	ENE	3,3	N	8,4	N	13,3	NNE	15,2	11,43
	9	NNE	13,2	N	9,3	N	8,2	NNW	5,6	N	16,8	N	11,3	10,73
	10	NNE	13,8	NNE	11,1	NE	8,7	NE	9,2	NNE	11,5	NNE	12,4	11.12
	11	NNE	13,4	N	22,2	NNE	26,2	NNE	38,8	NNE	31,9	N	19,3	25,30
	12	NNE	18,4	N	20,3	N	13,8	N	10,2	NNE	13,8	NNE	20,4	16,15
	13	NE	12,7	N	12,2	NNE	8,4	N	11,9	N	13,1	NNE	13,1	11.90
	14	N	14,2	N	13,2	N	6,2	NNW	9,1	N	11,2	N	10,7	10,77
	15	N	13,2	NNW	11,4	E	2,2	SSW	4,2	N	11,4	N	10,5	8,82
	16	N	12,0	N	12,2	NE	4,7	NNE	10,7	NNE	16,0	NNE	17,0	12,10
	17	NNE	13,4	N	12,1	ENE	2,6	N	7,4	N	12,1	NNE	8,5	9,35
	18	SE	4,0	SE	3,2	SE	5,0	N	4,2	N	6,4	S	7,3	5,02
	19	SE	8,2	SE	8,7	SE	13,3	SE	13,0	SE	10,6	SE	17,5	11,88
	20	SE	16,6	SE	17,1	SE	18,7	SE	16,5	SE	12,1	S	12,7	15,62
	21	SSE	13,7	SSE	12,2	W	5,6	W	4,1	WNW	6,4	NW	6,4	8,07
	22	E	3,6	NNE	9,2	SE	2,0	N	6,2	N	8,7	E	0,6	5,05
	23	SE	5,1	SE	7,6	SE	6,2	SE	2,7	SE	4,5	SE	3,6	4,95
	24	SE	2,7	NNE	5,0	NNE	2,7	NNE	3,2	N	7,8	SE	4,0	4,23
	25	SE	4,2	SE	3,9	SE	2,7	SE	8,6	SSE	2,7	SE	4,5	4,43
	26	SE	7,2	SE	3,4	SE	3,7	SE	3,6	E	2,1	ESE	11,8	5,20
	27	SE	11,0	NNE	2,6	ENE	5,7	NNE	3,7	NNW	11,4	N	16,8	8,53

Month & Year	Day	2h am		6h am		10h am		2h pm		6h pm		10h pm		Mean
YAMAGUCHI 35m.														
VIII 1889			mps		mps		mps		mps		mps		mps	mps
	1	N	1.8	NNE	0.7	S	3.8	SW	2.8	S	5.1	N	2.6	2.47
	2	—	0.0	—	0.2	SW	2.0	SW	2.7	SW	1.2	SW	0.8	1.15
	3	—	0.0	—	0.0	N	4.3	N	3.2	N	3.7	NE	1.8	2.17
	4	—	0.1	—	0.4	NNW	3.1	N	5.0	NNW	1.9	S	0.7	1.87
	5	N	0.7	N	0.5	S	1.6	SSW	3.4	SW	2.4	N	1.6	1.70
	6	N	1.3	—	0.2	WSW	1.7	S	3.3	NW	1.5	N	1.3	1.55
	7	N	0.5	—	0.1	SW	1.6	W	2.7	W	1.4	N	0.9	1.20
	8	NE	2.1	SE	1.2	SW	1.6	SSW	3.3	NW	1.3	N	0.9	1.73
	9	—	0.2	N	0.5	W	2.2	SW	2.5	NE	2.7	NE	1.7	1.62
	10	—	0.2	NNE	1.4	S	1.2	S	3.9	NNE	3.7	N	1.6	2.00
	11	N	1.1	NNW	1.5	SSW	2.4	NNW	2.5	NNE	3.3	—	0.2	1.83
	12	N	0.5	—	0.0	N	1.1	SSW	2.7	N	1.4	NNW	1.5	1.20
	13	N	1.7	N	0.9	WNW	1.7	W	1.7	SW	1.2	N	2.1	1.55
	14	N	1.1	N	2.7	SSE	5.7	SSW	7.5	SSW	3.5	N	2.2	3.78
	15	N	2.8	N	2.0	S	8.5	SSE	5.1	SSE	2.7	SW	2.7	3.97
	16	SE	3.2	NE	0.8	SE	5.4	S	5.7	SSE	2.4	S	2.8	3.38
	17	N	0.7	NNE	1.2	W	2.8	SSW	4.1	SW	4.2	NE	0.8	2.30
	18	—	0.4	N	0.5	NNW	1.2	N	6.0	N	9.9	N	5.2	3.87
	19	NNE	8.3	N	9.1	NNE	8.2	N	8.9	NNE	6.6	N	4.7	7.63
	20	N	1.7	SSW	1.8	ENE	1.8	N	6.5	NW	0.7	N	0.7	2.20
	21	N	0.8	—	0.2	S	1.3	SSW	3.3	W	0.9	N	1.9	1.40
	22	N	0.7	—	0.4	SSE	3.0	SSE	3.8	SSE	1.8	N	2.0	1.95
	23	N	0.7	NNW	0.5	SSE	4.1	SSE	5.7	SSW	2.8	E	2.0	2.63
	24	NNE	0.7	SE	1.8	SSW	5.9	SSE	4.4	S	5.6	SSE	1.3	3.28
	25	SE	3.1	S	3.7	SSE	5.2	S	1.7	—	0.2	—	0.1	2.33
	26	N	4.6	N	6.8	NNE	5.7	NNE	8.2	N	4.9	NNE	8.3	6.42
	27	N	6.4	N	5.2	N	4.4	N	6.8	NNE	5.7	—	0.2	4.78
	28	NNE	3.5	NE	4.2	NNW	6.7	NNW	6.2	NNE	3.2	N	2.6	4.40
	29	N	2.2	—	0.0	NNW	5.4	NNE	7.6	NE	3.8	NE	2.7	3.62
	30	N	1.2	NW	2.5	SSE	1.9	NNW	4.1	N	4.8	N	2.8	2.80
	31	—	0.4	NE	2.1	NE	2.2	NE	2.1	NE	2.5	NNE	1.2	1.75
	Mean		1.70		1.71		3.47		4.43		3.05		2.00	2.73
IX 1889	1	—	0.2	NNE	1.2	S	3.2	SSE	3.2	S	0.8	S	1.9	1.75
	2	N	2.0	N	1.9	N	1.1	N	3.9	N	2.7	NNE	2.4	2.33
	3	—	0.0	NE	3.4	S	2.3	SSW	5.2	N	3.2	N	2.2	2.72
	4	—	0.0	—	0.0	N	0.7	S	2.5	S	1.7	N	1.6	1.08
	5	NNE	0.7	—	0.0	NNW	0.7	S	1.5	S	3.9	N	0.2	1.17
	6	NNE	1.6	N	0.5	N	0.6	N	6.0	N	4.9	NNE	1.7	2.55
	7	—	0.2	—	0.0	NNW	2.6	N	4.9	N	4.4	—	0.1	2.03
	8	—	0.0	—	0.0	NE	0.7	WSW	2.6	NE	2.7	ENE	2.3	1.38
	9	SW	0.7	—	0.2	N	2.4	NNW	2.7	NNE	2.5	NNE	2.5	1.83
	10	NNE	2.8	SSE	0.7	N	5.0	N	4.7	NE	3.8	—	0.0	2.83
	11	NNE	4.7	NE	1.8	NNE	7.7	NNE	7.0	N	5.2	N	3.9	5.05
	12	N	2.3	—	0.2	NNE	4.7	NNE	7.7	NE	3.0	E	0.5	3.07
	13	—	0.0	—	0.0	N	3.5	N	6.3	NNE	3.7	N	1.2	2.45
	14	—	0.0	—	0.0	NNE	4.7	NNW	4.1	N	3.0	—	0.2	2.00
	15	—	0.0	—	0.1	S	1.1	N	1.7	N	1.7	—	0.2	0.80
	16	—	0.0	—	0.0	NE	3.4	N	5.8	NE	2.5	NE	2.2	2.32
	17	N	1.1	—	0.0	NE	0.9	NNE	3.9	NNE	2.4	NNE	1.2	1.58
	18	—	0.0	—	0.0	SW	1.2	SW	3.8	ENE	1.9	NNE	1.3	1.37
	19	NNE	1.2	—	0.0	ESE	2.2	S	3.4	N	2.1	NNE	3.1	2.00
	20	NNE	1.4	WSW	1.8	SW	2.7	SW	3.2	S	5.2	W	0.8	2.52
	21	—	0.0	SW	3.3	SSW	1.3	S	1.9	SSW	0.7	WNW	1.5	1.45
	22	N	1.8	NNE	2.5	SW	1.5	SW	1.5	N	1.3	N	1.5	1.68
	23	—	0.4	—	0.2	—	0.2	SW	0.9	NNW	1.1	N	1.1	0.65
	24	N	0.9	N	0.6	—	0.0	SSW	2.1	ENE	2.4	N	1.5	1.25
	25	—	0.2	—	0.2	S	0.5	S	4.3	WSW	1.7	N	1.5	1.40
	26	N	0.8	—	0.0	SSW	0.7	SW	2.4	N	1.2	NNE	2.5	1.27
	27	NNE	1.6	NNE	1.2	NE	1.6	SSW	1.2	SSW	1.1	SE	1.2	1.32

TABLE V. WIND DIRECTION AND WIND VELOCITY.

Month & Year	Day	2h am		6h am		10h am		2h pm		6h pm		10h pm		Mean
HIGASHI HOBEN 736m.														
IX 1889			mps		mps		mps		mps		mps		mps	mps
	28	NNE	9.1	N	11,9	N	10,5	N	12,6	N	13,1	NE	9,9	11.18
	29	N	11,0	N	9,6	N	11.2	N	7.2	N	8,0	N	8,7	9.28
	30	NNW	9,5	N	2.25	N	9.2	N	10,8	NE	11,8	N	12,5	12.72
	Mean		10,56		10,78		7,58		9,00		11,02		11,08	10,00
X 1889	1	N	10,8	N	8,7	N	4.1	NNW	6,5	N	11,4	N	8,2	8,28
	2	SSW	3,3	ESE	5,2	SSE	6,5	WNW	1,0	SE	0,9	SSW	5,6	3,75
	3	W	3,4	N	18,7	N	8,2	N	5,9	N	11,1	N	9,4	9,45
	4	N	12,1	N	14,2	NE	3,2	SE	2,7	N	11,1	N	6,7	8,33
	5	NE	4,7	S	4,2	SE	7,7	SE	14,6	SE	14,0	SE	10,6	9,20
	6	SSE	5,1	N	18,8	NW	3,5	N	9,6	N	12,5	NNE	7,3	9,47
	7	N	7,8	N	14,0	W	4,8	N	7,4	N	14,0	N	7,8	9,30
	8	NNE	7,8	NE	10,7	E	2,4	N	9,4	N	15,1	N	12,2	9,60
	9	NNE	13,8	NNE	15,1	NNE	9,4	NNE	5,7	NNE	9,3	E	2,2	9,25
	10	S	2,5	NW	7,7	NE	3,4	NE	5,3	NNE	11,2	NNE	9,2	6,55
	11	NE	6,4	ESE	4,2	ENE	2,2	SE	4,6	SSE	1,0	SE	2,1	3,42
	12	ESE	7,2	NE	9,6	NE	4,9	SE	2,0	N	15,9	NE	19,4	9,83
	13	NE	14,5	NE	18,6	NE	11,3	NE	8,9	NE	15,7	ENE	10,8	13,30
	14	ENE	9,8	ESE	11,8	ESE	7,7	SE	13,3	SE	12,2	SE	9,2	10,67
	15	SE	13,1	NNE	12,2	SE	11,4	SE	16,3	E	5,3	N	16,8	12,52
	16	N	23,0	N	16,7	N	20,2	N	14,5	N	13,2	N	12,5	16,68
	17	N	13,4	N	8,2	NE	2,9	N	1,5	N	10,1	N	7,7	7,30
	18	N	6,7	N	12,0	NE	3,7	NE	1,9	NNE	9,7	NNE	7,7	6,95
	19	SE	8,1	ESE	5,5	SE	2,2	SE	9,3	ESE	16,9	SE	15,9	9,65
	20	SE	17,5	SE	18,2	SE	23,1	SE	22,0	SE	22,6	SE	21,9	20,88
	21	SE	16,8	SW	14,1	W	11,3	NW	14,8	W	9,0	WNW	13,3	13,22
	22	NW	12,7	NNW	13,1	N	12,7	NW	11,5	N	9,0	N	8,3	11,22
	23	N	7,6	N	7,2	NNE	7,1	NW	7,3	NW	5,2	W	4,2	6,43
	24	W	3,4	WSW	5,4	NW	1,6	NW	2,2	N	8,3	N	6,9	4,63
	25	N	10,0	NNE	9,9	ENE	4,1	SE	4,2	N	9,6	N	10,9	8,12
	26	N	12,2	NNW	4,2	N	2,5	N	2,2	N	2,4	SE	9,9	5,57
	27	SE	7,8	SE	7,6	SE	7,4	SE	7,6	SE	14,9	SE	15,7	10,17
	28	SE	15,2	ESE	18,6	N	11,9	NNE	11,1	N	11,0	NW	9,8	12,93
	29	W	8,7	N	17,1	N	24,7	N	15,0	NW	7,7	N	14,6	14,63
	30	N	12,0	NNW	8,9	WNW	5,7	WSW	2,4	W	3,2	WSW	7,7	6,65
	31	W	7,5	W	4,8	WSW	5,6	SSE	5,7	SW	3,2	W	3,9	5,12
	Mean		9,83		11.13		7,06		7,95		10,22		9,96	9,46

Month & Year	Day	2h am		6h am		10h am		2h pm		6h pm		10h pm		Mean
YAMAGUCHI 35m.														
			mps		mps		mps		mps		mps		mps	mps
IX 1889	28	—	0.2	—	0.1	N	0.7	NNE	5.2	N	2.2	N	1.2	1.60
	29	—	0.2	—	0.4	NE	4.3	NNW	6.0	NNE	1.7	N	0.7	2.22
	30	—	0.1	NE	3.7	NNE	5.0	NNW	4.4	N	2.1	ENE	0.5	2.63
	Mean		0.84		0.80		2.24		3.80		2.56		1.42	1.94
X 1889	1	—	0.2	N	0.7	NNW	3.0	NNW	2.5	NNW	4.3	—	0.2	1.82
	2	N	1.2	N	1.2	SSW	0.7	SSW	0.9	N	1.8	—	0.4	1.03
	3	—	0.0	—	0.2	NE	4.9	SSW	4.6	N	1.7	SSW	0.6	2.00
	4	—	0.0	—	0.1	S	1.3	WSW	3.5	ENE	2.3	N	1.6	1.47
	5	S	0.7	N	1.4	—	0.2	S	6.2	N	0.9	N	2.2	1.93
	6	—	0.0	—	0.4	SSW	1.8	ENE	0.9	N	1.8	NE	1.8	1.12
	7	—	0.2	—	0.1	S	2.0	N	4.9	NE	2.7	N	0.7	1.77
	8	—	0.2	—	0.0	NE	0.7	N	2.3	NNW	5.8	—	0.2	1.53
	9	N	0.7	N	1.7	NNE	2.4	N	5.0	—	0.4	—	0.2	1.73
	10	—	0.4	N	0.9	—	0.4	SSW	2.3	N	0.8	N	1.4	1.03
	11	—	0.4	N	0.5	N	0.8	S	3.1	N	1.1	N	0.8	1.12
	12	N	1.1	ENE	1.9	SE	2.7	WSW	2.7	N	4.7	—	0.1	2.25
	13	—	0.2	—	0.0	NNE	4.5	NNE	8.5	NNE	3.5	N	2.3	3.17
	14	—	0.1	N	1.7	NE	3.2	S	4.7	N	2.4	—	0.4	2.08
	15	—	0.9	ENE	2.2	—	0.2	N	1.1	N	1.2	S	0.8	1.07
	16	NNW	1.6	N	5.4	N	4.9	NNW	6.6	N	1.2	—	0.2	3.32
	17	—	0.4	—	0.2	S	2.2	SSW	2.4	N	6.6	SSE	0.6	2.07
	18	—	0.0	N	0.5	E	1.8	SW	2.7	NE	1.2	N	0.9	1.18
	19	—	0.1	—	0.2	NE	1.7	E	1.6	N	3.2	N	3.2	1.67
	20	N	1.1	N	2.2	SSE	5.6	SSE	5.0	NW	2.1	SSW	3.9	3.32
	21	SSW	4.4	—	0.1	SW	4.7	W	4.6	SW	1.6	NNE	0.8	2.70
	22	N	0.8	N	0.8	WNW	3.9	N	5.1	N	3.0	N	2.7	2.72
	23	N	2.4	—	0.0	SSE	1.8	SW	3.9	—	0.2	—	0.2	1.27
	24	—	0.0	—	0.1	—	0.4	SW	3.8	N	1.2	NE	0.7	1.03
	25	—	0.4	—	0.0	S	0.7	ESE	2.1	N	1.4	—	0.4	0.83
	26	SSW	0.7	—	0.2	NNW	0.7	N	1.8	N	3.0	N	1.2	1.27
	27	—	0.1	N	1.2	SSE	0.6	S	3.7	N	2.6	N	1.2	1.57
	28	N	1.2	N	0.7	—	0.4	N	4.6	N	0.8	NNE	0.7	1.40
	29	N	2.7	NNW	2.2	N	9.5	N	7.7	NNW	0.9	S	1.9	4.15
	30	N	1.1	ENE	0.5	SSW	0.7	SSW	2.0	NNW	2.6	—	0.1	1.17
	31	NW	0.8	—	0.2	—	0.0	SW	1.6	NE	1.9	N	1.4	0.98
	Mean		0.78		0.89		2.21		3.60		2.22		1.10	1.80

TABLE VI. AMOUNT OF PRECIPITATION.

Month & Year	Day	2h am	4h am	6h am	8h am	10h am	Noon	2h pm	4h pm	6h pm	8h pm	10h pm	M.N.	Sum
FUJI 3718m.														
VIII 1889		mm	mm	mm	mm	mm	mm	mm	mm	mm	mm	mm	mm	mm
	1	—	—	—	—	—	—	—	—	—	—	—	—	—
	2	—	—	—	—	—	—	—	—	—	—	—	—	—
	3	—	—	—	—	—	—	—	—	—	—	—	—	—
	4	—	—	—	—	—	—	—	—	—	—	—	—	—
	5	—	—	—	—	—	—	—	—	—	—	—	—	—
	6	—	—	—	—	—	—	—	—	—	—	—	—	—
	7	—	—	—	—	—	—	—	—	0.0	—	—	—	0.0
	8	—	—	—	—	—	—	—	—	7.8	3.4	6.2	4.1	21.5
	9	0.5	2.6	0.5	—	—	—	—	—	0.4	—	—	—	4.0
	10	—	—	—	—	—	—	—	—	—	1.3	—	—	1.3
	11	—	—	—	—	—	—	—	—	—	—	—	—	—
	12	—	—	—	—	—	—	—	—	—	—	—	—	—
	13	—	—	—	—	—	—	—	1.5	—	—	—	—	1.5
	14	—	—	—	—	—	—	—	0.1	—	—	—	—	0.1
	15	—	—	—	—	—	—	—	—	—	—	—	—	—
	16	—	—	—	—	—	—	—	—	—	—	—	—	—
	17	—	—	—	—	—	—	—	—	—	0.3	—	—	0.3
	18	—	—	1.4	0.2	0.3	8.4	13.8	5.0	13.3	9.0	21.7	69.2	142.3
	19	25.9	27.6	5.3	15.9	39.4	16.5	3.7	4.4	4.5	3.7	9.6	9.6	176.1
	20	9.6	9.6	9.6	9.6	53.2	44.1	36.2	36.3	27.7	27.7	27.7	27.7	319.0
	21	27.7	27.7	27.6	14.5	0.2	—	—	—	—	—	—	—	97.7
	22	—	—	—	—	—	—	—	—	—	—	—	—	—
	23	—	—	—	—	—	—	—	0.1	6.0	2.5	0.5	—	9.1
	24	—	—	—	—	—	—	—	—	3.5	—	—	—	3.5
	25	—	—	—	—	—	—	—	—	—	—	—	—	—
	26	—	—	—	—	—	—	—	0.4	14.6	7.6	0.0	—	22.6
	27	3.8	—	—	—	4.0	6.0	0.9	6.9	3.3	—	—	—	24.9
	28	11.2	0.6	—	—	—	0.4	3.9	9.3	15.2	1.5	1.6	0.7	44.4
	29	—	3.3	3.6	0.7	0.5	3.1	5.4	1.5	0.6	0.6	0.1	—	19.4
	30	—	—	0.4	—	—	—	—	—	—	—	—	—	0.4
	31	—	—	—	—	—	—	—	—	—	—	—	—	—
	Sum	78.7	84.4	48.4	50.9	97.6	78.5	63.9	65.5	96.9	57.6	67.4	111.3	888.1
IX 1889	1	—	..	—	..	—	..	—	..	2.9	..	—	..	2.9
	2	—	..	13.5	..	7.4	..	16.6	..	10.4	..	—	..	47.9
	3	—	..	79.2	..	—	..	4.3	..	—	..	—	..	83.5
	4	—	..	—	..	—	..	—	..	—	..	—	..	—
	5	—	..	5.6	..	—	..	—	..	—	..	—	..	5.6
	6	—	..	32.3	..	11.7	..	1.4	..	—	..	—	..	45.4
	7	—	..	—	..	—	..	—	..	—	..	—	..	—
	Sum	—	..	179.0	..	116.7	..	86.2	..	110.2	..	—	..	1073.4
ONTAKE 3062m.														
VIII 1891	1	0.2	..	—	..	—	..	—	..	—	..	0.5	..	0.7
	2	1.4	..	1.4	..	4.8	..	43.4	..	12.9	..	0.4	..	64.3
	3	2.9	..	4.0	..	0.7	..	2.0	..	—	..	1.3	..	10.9
	4	2.8	..	15.4	..	32.9	..	0.9	..	6.5	..	—	..	58.5
	5	3.4	..	0.9	..	0.8	..	0.0	..	17.2	..	2.2	..	24.5
	6	0.8	..	4.7	..	0.7	..	8.7	..	2.0	..	—	..	16.9
	7	0.4	..	0.9	..	—	..	—	..	1.4	..	0.2	..	2.9
	8	—	..	—	..	—	..	—	..	—	..	—	..	—
	9	—	..	—	..	—	..	—	..	—	..	—	..	—
	10	0.6	..	0.6	..	0.2	..	—	..	5.7	..	0.1	..	7.2
	11	0.7	..	—	..	0.9	..	1.3	..	0.0	..	0.1	..	3.0
	12	0.7	..	—	..	—	..	—	..	—	..	—	..	0.7
	13	1.2	..	1.0	..	—	..	—	..	—	..	—	..	2.2
	14	—	..	—	..	—	..	—	..	—	..	0.5	..	0.5
	15	—	..	—	..	—	..	—	..	—	..	—	..	—
	16	0.2	..	0.1	..	—	..	—	..	0.7	..	2.7	..	3.7
	17	25.9	..	26.0	..	26.0	..	6.7	..	6.1	..	15.5	..	106.2

TABLE VI*. AMOUNT OF PRECIPITATION

Month & Year	Day	2h am	4h am	6h am	8h am	10h am	Noon	2h pm	4h pm	6h pm	8h pm	10h pm	M.N.	Sum
YAMANAKA 990m.														
VIII 1889		mm	mm	mm	mm	mm	mm	mm	mm	mm	mm	mm	mm	mm
	1	—	..	—	—	—	—	—	—	—	—	—	—	—
	2	—	..	—	—	—	—	—	0.1	—	—	—	—	0.1
	3	—	..	—	—	—	—	—	—	—	—	—	—	—
	4	—	..	—	—	—	—	—	—	—	—	—	—	—
	5	—	..	—	—	—	—	—	—	—	—	—	—	—
	6	—	..	—	—	—	—	—	—	—	—	—	—	—
	7	—	..	—	—	—	—	—	0.0	—	—	—	—	0.0
	8	—	..	—	—	—	—	—	—	1.8	0.0	0.2	—	2.0
	9	—	..	—	—	—	—	—	—	—	—	—	—	—
	10	—	..	—	—	—	—	—	—	—	0.0	—	—	0.0
	11	0.0	..	—	—	—	—	—	—	—	—	0.0	—	0.0
	12	—	..	—	—	—	—	—	—	—	—	—	—	—
	13	—	..	—	—	—	—	—	—	—	—	—	—	—
	14	—	..	—	—	—	—	—	—	—	—	—	—	—
	15	—	..	—	—	—	—	—	—	—	—	—	—	—
	16	—	..	—	—	—	—	—	—	—	—	—	—	—
	17	—	..	—	—	—	—	—	—	0.0	0.0	—	—	0.0
	18	—	..	—	0.5	0.1	1.7	13.5	22.0	15.2	20.1	80.0	51.1	204.2
	19	4.4	..	23.1	0.7	1.3	20.4	6.1	2.6	7.8	0.8	—	—	67.2
	20	0.4	..	15.9	21.0	23.0	24.9	27.8	11.4	6.6	15.4	11.3	7.0	165.6
	21	6.2	..	18.0	0.0	—	0.0	—	—	—	—	—	—	24.2
	22	—	..	—	—	—	—	—	—	—	—	—	—	—
	23	—	..	—	—	—	—	—	3.9	18.4	3.5	0.6	0.0	26.4
	24	—	..	—	—	—	—	—	—	—	—	—	—	—
	25	—	..	—	—	—	—	—	0.0	0.1	0.3	0.0	0.0	0.4
	26	0.0	..	—	—	—	—	—	0.0	4.0	6.2	0.9	0.0	11.1
	27	—	..	0.1	0.2	6.6	6.5	3.7	2.4	0.2	0.7	0.4	0.9	22.1
	28	15.0	..	7.6	0.2	0.1	2.6	2.8	4.7	7.9	3.2	0.6	0.4	45.2
	29	0.5	..	7.0	0.2	0.6	1.5	0.2	0.3	0.1	0.0	0.0	0.5	10.9
	30	0.8	..	0.1	0.0	0.0	—	—	—	—	—	—	0.0	0.9
	31	—	..	0.1	—	—	—	—	—	—	—	—	—	0.1
	Sum	27.1	..	71.9	22.9	31.7	55.0	4.1	47.4	62.1	50.2	94.0	60.5	580.1
IX 1889	1	—	..	—	..	—	..	0.0	..	0.0	..	0.0	..	0.0
	2	0.0	..	0.2	..	11.2	..	7.2	..	0.1	..	0.1	..	18.8
	3	3.6	..	3.4	..	1.5	..	0.7	..	—	..	—	..	9.2
	4	—	..	..	..	—	..	—	..	—	..	—	..	—
	5	—	..	1.5	..	—	..	0.0	..	0.2	..	1.2	..	2.9
	6	2.2	..	2.2	..	0.4	..	—	..	—	..	—	..	4.8
	7	—	..	0.1	..	—	..	—	..	—	..	—	..	0.1
	Sum	33.1	..	79.1	..	44.8	..	62.0	..	62.4	..	95.3	..	596.2
KUROSAWA 832m.														
VIII 1891	1	—	—	—	—	—	—	—	—	—	—	—	—	—
	2	—	—	—	0.0	0.0	2.2	5.0	3.5	0.6	—	—	—	11.3
	3	0.0	—	0.0	—	0.2	0.0	0.1	0.2	—	—	0.1	—	0.4
	4	—	0.0	1.4	0.1	1.1	3.7	—	0.2	1.0	—	—	0.0	7.5
	5	0.1	6.2	4.7	0.0	0.3	0.0	0.2	0.0	0.2	0.9	0.6	0.3	13.5
	6	0.0	0.1	0.4	0.4	—	0.5	1.3	—	1.0	—	—	—	3.7
	7	0.0	—	—	—	—	—	0.0	1.6	0.2	—	—	—	1.8
	8	—	—	—	—	—	—	—	—	—	—	—	—	—
	9	—	—	—	—	—	—	—	—	—	—	—	—	—
	10	—	—	—	0.0	0.0	0.0	—	0.0	—	2.1	—	—	2.1
	11	—	—	—	—	—	—	—	—	—	—	—	—	—
	12	—	—	—	—	—	—	—	—	—	—	—	—	—
	13	—	—	—	—	—	—	—	—	—	—	—	—	—
	14	—	—	—	—	—	—	—	—	—	—	—	—	—
	15	—	—	—	—	—	—	—	—	—	—	—	—	—
	16	—	—	—	—	—	—	—	—	0.2	0.5	3.0	15.7	19.4
	17	8.9	17.2	14.2	38.7	5.4	2.1	0.0	0.2	7.5	0.0	10.7	4.1	109.5

TABLE VI. AMOUNT OF PRECIPITATION.

Month & Year	Day	2h am	4h am	6h am	8h am	10h am	Noon	2h pm	4h pm	6h pm	8h pm	10h pm	M.N.	Sum
ONTAKE 3062m.														
VIII 1891		mm	mm	mm	mm	mm	mm	mm	mm	mm	mm	mm	mm	mm
	18	9.5	..	19.3	..	9.9	..	14.5	..	8.4	..	—	..	61.6
	19	—	..	—	..	—	..	—	..	—	..	—	..	—
	20	—	..	—	..	—	..	—	..	—	..	—	..	—
	21	—	..	0.4	..	10.2	..	40.7	..	22.2	..	36.1	..	109.6
	22	58.9	..	14.5	..	3.3	..	6.4	..	7.5	..	8.1	..	98.7
	23	6.7	..	4.4	..	7.9	..	5.0	..	4.1	..	—	..	28.1
	24	—	..	—	..	—	..	—	..	0.0	..	—	..	0.0
	25	—	..	—	..	—	..	—	..	—	..	—	..	—
	26	—	..	—	..	—	..	—	..	—	..	—	..	—
	27	—	..	—	..	—	..	—	..	—	..	—	..	—
	28	—	..	—	..	—	..	—	..	—	..	—	..	—
	29	—	..	—	..	—	..	—	..	—	..	—	..	—
	30	—	..	—	..	—	..	—	..	—	..	—	..	—
	31	—	..	—	..	—	..	—	..	—	..	—	..	—
	Sum	106.2	..	93.6	..	98.3	..	129.6	..	94.7	..	67.7	..	600.2
IX 1891	1	—	..	—	..	—	..	—	..	0.0	..	—	..	0.0
	2	—	..	0.4	..	—	..	—	..	0.0	..	—	..	0.4
	3	—	..	—	..	—	..	0.0	..	—	..	—	..	0.0
	4	—	..	—	..	—	..	—	..	—	..	—	..	—
	5	—	..	—	..	—	..	—	..	—	..	—	..	—
	6	—	..	—	..	—	..	—	..	—	..	—	..	—
	7	—	..	—	..	—	..	0.0	..	0.0	..	—	..	0.0
	8	0.0	..	0.7	..	—	..	1.2	..	0.1	..	2.5	..	4.4
	9	19.0	..	50.2	..	49.8	..	15.0	..	3.6	..	4.1	..	141.7
	10	2.4	..	1.4	..	0.2	..	0.5	..	2.7	..	7.4	..	14.6
	11	3.8	..	1.5	..	0.8	..	—	..	—	..	0.2	..	6.3
	12	0.4	..	—	..	—	..	—	..	—	..	0.2	..	0.6
	Sum	141.9	..	117.4	..	169.1	..	188.2	..	101.4	..	82.1	..	788.2
GOZAISHODAKE 1209m.														
IX 1888	4	0.0	20.6	1.6	—	—	1.0	0.0	0.0	0.9	2.6	—	—	26.7
	5	0.0	2.8	—	—	—	—	—	—	—	—	—	—	2.8
	6	—	—	—	—	—	—	—	—	—	—	—	—	—
	7	—	—	—	0.0	0.0	—	—	—	—	—	—	—	0.0
	8	—	—	—	0.0	0.0	1.0	0.0	0.0	—	—	—	0.0	1.0
	9	—	—	—	—	—	0.5	0.3	5.6	5.0	0.3	—	0.2	11.9
	10	—	—	0.2	2.9	13.0	2.0	4.0	0.2	1.3	0.0	20.6	18.4	58.3
	11	8.4	10.7	2.4	2.0	4.1	3.4	6.5	2.1	4.9	—	0.4	1.9	44.3
	12	18.0	41.2	12.0	1.2	3.5	4.9	1.2	2.3	1.4	0.9	2.0	0.2	94.1
	13	—	1.2	0.6	—	—	—	—	—	—	0.0	—	—	3.0
	14	—	0.5	—	—	—	—	—	—	—	—	—	—	0.5
	15	—	—	—	—	—	—	0.1	—	—	—	—	0.8	0.8
	16	—	—	1.1	0.9	0.1	—	0.1	—	—	—	—	—	2.2
	17	—	—	—	—	—	—	—	—	—	—	—	—	—
	18	—	—	—	—	—	—	—	—	—	—	—	—	—
	19	—	—	—	—	—	—	—	—	—	1.0	6.9	10.5	18.4
	20	93.9	42.6	27.5	31.8	26.6	8.1	3.1	1.2	1.3	0.3	0.2	4.7	177.9
	21	4.5	2.9	1.8	0.8	1.7	1.5	0.0	—	—	—	—	0.0	13.2
	22	0.5	1.3	0.7	—	0.2	0.7	1.7	4.0	3.2	0.2	—	—	12.5
	23	—	—	0.0	—	—	—	—	—	—	—	—	—	0.0
	24	—	—	—	—	—	—	—	—	—	—	—	—	—
	25	—	—	—	—	—	—	—	—	—	0.0	0.1	1.2	1.3
	26	0.2	—	0.1	—	—	—	—	—	—	—	—	—	0.3
	27	—	—	0.2	—	—	—	—	—	—	—	—	—	0.2
	28	—	—	—	—	—	—	—	—	—	—	—	—	—
	29	—	—	—	—	—	—	0.0	—	—	—	—	—	0.0
	30	—	—	—	—	—	—	—	—	—	—	—	—	—
X 1888	1	—	—	0.0	0.0	—	—	—	—	—	—	—	—	0.0
	2	0.0	—	—	—	—	—	—	—	—	—	—	—	0.0
	3	—	—	—	—	—	—	—	—	—	—	—	—	—
	Sum	62.5	123.6	48.2	39.6	49.2	23.1	17.1	15.4	18.0	5.1	29.6	37.6	509.1

TABLE VI^a. AMOUNT OF PRECIPITATION.

Month & Year	Day	2h am	4h am	6h am	8h am	10h am	Noon	2h pm	4h pm	6h pm	8h pm	10h pm	M.N.	Sum
KUROSAWA 832m.														
VIII 1894		mm	mm	mm	mm	mm	mm	mm	mm	mm	mm	mm	mm	mm
	18	0.7	1.6	0.9	2.1	8.1	—	8.4	0.7	4.5	—	—	—	26.7
	19	—	—	—	—	—	—	—	—	—	—	—	—	—
	20	—	—	—	—	—	—	—	—	—	—	—	—	—
	21	—	—	—	0.3	1.9	0.5	6.8	6.9	2.4	0.9	6.0	11.2	36.9
	22	9.6	14.2	0.5	—	0.2	1.7	0.1	0.0	0.4	1.6	0.1	1.7	30.1
	23	2.4	0.0	0.2	1.1	1.6	0.4	3.1	2.2	0.3	—	—	—	11.3
	24	—	—	—	—	—	—	—	—	0.0	—	—	0.0	0.0
	25	—	—	—	—	—	—	—	—	—	—	—	—	—
	26	—	—	—	—	—	—	—	—	—	—	—	—	—
	27	—	—	—	—	—	—	—	—	—	—	—	—	—
	28	—	—	—	—	—	—	—	—	—	—	—	—	—
	29	—	—	—	—	—	—	—	—	—	—	—	—	—
	30	—	—	—	—	—	—	—	—	—	—	—	—	—
	31	—	—	—	—	—	—	—	—	—	—	—	—	—
	Sum	21.7	29.3	22.3	12.7	12.5	11.1	25.8	15.3	14.5	6.5	20.5	30.0	260.2
IX 1894	1	—	—	—	—	—	—	—	—	—	—	—	—	—
	2	—	—	—	—	—	—	0.0	0.0	—	—	—	—	0.0
	3	—	—	—	—	—	—	—	—	—	—	—	—	—
	4	—	—	—	—	—	—	—	—	—	—	—	—	—
	5	—	—	—	—	—	—	—	—	—	—	—	—	—
	6	—	—	—	—	—	—	—	—	—	—	—	—	—
	7	—	—	—	—	—	—	—	—	—	0.0	—	—	0.0
	8	—	0.0	—	—	—	—	—	—	0.2	0.3	1.1	3.0	4.6
	9	8.9	7.3	8.5	6.6	17.1	4.1	0.0	0.4	—	0.7	0.2	—	53.8
	10	—	—	—	—	—	—	—	0.0	—	1.0	—	—	1.0
	11	—	—	—	—	—	—	—	—	—	—	—	—	—
	12	—	—	—	—	—	—	—	—	—	—	0.0	0.1	0.1
	Sum	30.6	16.6	30.8	19.3	29.6	15.2	25.9	15.7	18.5	8.5	21.8	35.1	327.5
YOKKAICHI 4m.														
IX 1888	4	0.0	..	2.8	..	—	..	0.0	..	—	..	0.2	..	3.0
	5	—	..	—	..	—	..	—	..	—	..	0.0	..	0.0
	6	—	..	—	..	—	..	—	..	—	..	—	..	—
	7	—	..	—	..	—	..	—	..	—	..	—	..	—
	8	—	..	—	..	—	..	0.0	..	—	..	—	..	0.0
	9	—	..	—	..	—	..	0.0	..	—	..	—	..	0.0
	10	—	..	0.8	..	22.6	..	9.2	..	1.1	..	5.0	..	38.7
	11	4.2	..	4.8	..	1.8	..	0.7	..	0.0	..	—	..	11.5
	12	0.0	..	12.8	..	8.4	..	0.0	..	—	..	0.4	..	21.6
	13	—	..	—	..	—	..	0.0	..	—	..	—	..	0.0
	14	—	..	—	..	—	..	—	..	—	..	—	..	—
	15	—	..	—	..	—	..	—	..	—	..	0.0	..	0.0
	16	0.0	..	0.4	..	1.9	..	0.0	..	0.6	..	0.0	..	2.9
	17	—	..	—	..	—	..	—	..	—	..	—	..	—
	18	—	..	—	..	—	..	—	..	—	..	—	..	—
	19	—	..	—	..	—	..	—	..	—	..	—	..	—
	20	1.1	..	11.9	..	3.2	..	0.0	..	0.0	..	0.0	..	16.2
	21	13.8	..	0.3	..	5.0	..	0.9	..	0.6	..	—	..	20.1
	22	0.2	..	0.2	..	0.0	..	2.0	..	5.1	..	—	..	7.5
	23	—	..	—	..	—	..	—	..	—	..	—	..	—
	24	—	..	—	..	—	..	—	..	—	..	—	..	—
	25	—	..	—	..	—	..	—	..	—	..	0.0	..	0.0
	26	0.0	..	—	..	—	..	—	..	—	..	—	..	0.0
	27	—	..	—	..	—	..	—	..	—	..	—	..	—
	28	—	..	—	..	—	..	—	..	—	..	—	..	—
	29	—	..	—	..	—	..	—	..	—	..	—	..	—
	30	—	..	—	..	—	..	—	..	—	..	—	..	—
X 1888	1	—	..	—	..	—	..	—	..	—	..	—	..	—
	2	—	..	—	..	—	..	—	..	—	..	—	..	—
	3	—	..	—	..	—	..	—	..	—	..	—	..	—
	Sum	18.8	..	31.0	..	48.2	..	12.8	..	7.4	..	5.6	..	121.8

TABLE VI. AMOUNT OF PRECIPITATION.

Month & Year	Day	2^{h} a.m.	6^{h} a.m.	10^{h} a.m.	2^{h} p.m.	6^{h} p.m.	10^{h} p.m.	Sum
HIGASHI HOBEN 736m.								
		mm	mm	mm	mm	mm	mm	mm
VIII 1889	1	0,0	0,2	0,0	—	—	0,0	0,2
	2	0,1	0,2	0,0	0,0	0,0	—	0,3
	3	—	—	—	—	—	—	—
	4	—	—	—	—	—	—	—
	5	—	—	—	—	—	—	—
	6	—	0,0	0,0	—	—	0,0	0,0
	7	0,0	0,1	0,0	—	—	—	0,1
	8	0,0	0,1	0,0	—	—	—	0,1
	9	—	—	—	—	—	—	—
	10	—	—	—	—	—	—	—
	11	—	—	—	—	—	0,0	0,0
	12	0,0	—	—	—	—	—	0,0
	13	—	—	0,0	—	—	0,0	0,0
	14	0,0	0,0	0,0	0,0	0,0	0,0	0,0
	15	0,0	0,1	0,1	—	0,0	—	0,2
	16	0,0	0,1	0,0	—	—	—	0,1
	17	0,0	0,1	0,0	—	—	—	0,1
	18	—	—	—	—	0,0	—	0,0
	19	—	—	0,0	0,2	0,1	0,2	0,5
	20	0,1	2,7	0,7	0,4	0,1	0,0	4,0
	21	0,1	0,1	0,0	—	0,0	—	0,2
	22	—	0,0	0,0	—	0,0	—	0,0
	23	0,0	0,3	0,0	—	—	0,0	0,3
	24	0,2	0,6	0,2	0,0	0,0	0,1	1,1
	25	0,7	0,6	0,4	0,7	5,4	2,3	10,1
	26	5,0	0,3	0,1	0,0	0,0	0,0	5,4
	27	0,1	0,2	0,0	0,0	—	—	0,3
	28	—	—	—	—	—	—	—
	29	—	—	—	—	—	—	—
	30	—	—	—	—	—	—	—
	31	—	0,0	0,2	2,1	0,2	0,1	2,6
	Sum	6,3	5,7	1,7	3,4	5,8	2,7	25,6
IX 1880	1	—	0,0	0,0	—	0,0	0,1	0,1
	2	0,2	0,1	39,6	9,5	0,0	0,0	49,4
	3	0,0	0,1	—	—	—	—	0,1
	4	—	—	—	—	—	—	—
	5	0,0	0,0	0,0	0,0	0,0	0,0	0,0
	6	4,9	8,1	0,0	0,1	—	0,0	13,1
	7	—	—	—	—	—	0,0	0,0
	8	—	0,0	—	—	—	—	0,0
	9	—	—	—	—	—	—	—
	10	—	—	—	—	—	1,3	1,3
	11	0,2	0,1	0,0	0,0	0,0	0,7	1,0
	12	0,7	0,5	0,0	0,0	—	—	1,2
	13	—	—	—	—	—	—	—
	14	—	—	—	—	—	—	—
	15	0,0	—	—	0,1	0,0	—	0,1
	16	—	—	—	—	—	—	—
	17	—	—	—	—	—	—	—
	18	—	—	—	—	—	—	—
	19	—	—	—	—	0,1	2,3	2,4
	20	0,9	0,6	0,4	0,5	0,7	1,6	5,1
	21	2,4	12,6	5,9	3,0	0,1	—	24,0
	22	0,0	0,0	—	—	—	—	0,0
	23	—	—	—	—	—	—	—
	24	—	—	—	—	—	—	—
	25	—	—	—	—	—	—	—
	26	0,1	—	—	—	—	—	0,1
	27	0,0	2,1	1,2	0,3	1,0	4,1	4,7

TABLE VIa. AMOUNT OF PRECIPITATION

Month & Year	Day	2^{h} a.m.	6^{h} a.m.	10^{h} a.m.	2^{h} p.m.	6^{h} p.m.	10^{h} p.m.	Sum
YAMAGUCHI 35m.								
		mm	mm	mm	mm	mm	mm	mm
VIII 1889	1	—	—	—	—	—	—	—
	2	—	—	—	—	—	—	—
	3	—	—	—	—	—	—	—
	4	—	—	—	—	—	—	—
	5	—	—	—	—	—	—	—
	6	—	—	—	—	—	—	—
	7	—	—	—	—	—	—	—
	8	—	—	—	—	—	—	—
	9	—	—	—	—	—	—	—
	10	—	—	—	—	—	—	—
	11	—	—	—	—	—	—	—
	12	—	—	—	—	—	—	—
	13	—	—	—	—	—	—	—
	14	—	—	—	—	—	—	—
	15	—	0,3	0,0	—	—	—	0,3
	16	—	—	—	—	—	—	—
	17	—	—	0,0	—	—	—	0,0
	18	—	—	—	—	0,0	—	0,0
	19	—	—	0,0	0,0	0,4	0,1	0,5
	20	1,8	2,0	1,1	0,3	0,0	—	5,2
	21	—	—	—	—	—	—	—
	22	—	—	—	—	—	—	—
	23	—	—	—	—	—	0,0	0,0
	24	—	—	0,0	—	—	—	0,0
	25	1,8	0,1	0,5	0,7	2,0	0,2	5,3
	26	5,6	—	0,0	—	—	—	5,6
	27	—	—	—	—	—	—	—
	28	—	—	—	—	—	—	—
	29	—	—	—	—	—	—	—
	30	—	—	—	—	—	—	—
	31	—	—	0,1	2,9	0,2	0,0	3,2
	Sum	9,2	2,4	1,7	3,9	8,3	0,5	26,0
IX 1889	1	—	—	—	—	—	—	—
	2	—	—	62,4	14,9	—	—	77,3
	3	—	—	—	—	—	—	—
	4	—	—	—	—	—	—	—
	5	—	—	—	—	—	—	—
	6	1,3	8,1	0,0	—	—	—	9,4
	7	—	—	—	—	—	—	—
	8	—	—	—	—	—	—	—
	9	—	—	—	—	—	—	—
	10	—	—	—	—	—	4,5	4,5
	11	0,3	—	0,0	0,0	0,0	0,8	1,1
	12	0,7	0,1	—	—	—	—	0,8
	13	—	—	—	—	—	—	—
	14	—	—	—	—	—	—	—
	15	—	—	—	0,0	0,0	—	0,0
	16	—	—	—	—	—	—	—
	17	—	—	—	—	—	—	—
	18	—	—	—	—	—	—	—
	19	—	—	—	—	0,3	5,1	5,4
	20	2,0	1,4	0,0	1,0	1,3	2,4	8,1
	21	3,0	21,9	14,1	2,2	1,6	—	42,8
	22	—	—	—	—	—	—	—
	23	—	—	—	—	—	—	—
	24	—	—	—	—	—	—	—
	25	—	—	—	—	—	—	—
	26	—	—	—	—	—	—	—
	27	—	1,2	1,9	0,0	1,5	0,0	5,6

TABLE VI. AMOUNT OF PRECIPITATION.

Month & Year	Day	2h am	6h am	10h am	2h pm	6h pm	10h pm	Sum
HIGASHI HOBEN 736m.								
		mm	mm	mm	mm	mm	mm	mm
IX 1889	28	0,0	0,0	0,4	0,0	0,0	0,0	0,4
	29	0,0	0,0	0,0	0,0	—	—	0,0
	30	—	0,0	—	—	—	—	0,0
	Sum	9,4	24,2	47,5	13,9	1,9	6,4	103,9
X 1889	1	—	—	—	—	—	—	—
	2	—	—	—	—	—	—	—
	3	—	—	—	—	—	—	—
	4	—	—	—	—	—	—	—
	5	—	—	—	—	—	—	—
	6	2,6	8,8	1,7	0,2	0,0	0,0	13,3
	7	0,0	0,0	—	—	—	—	0,0
	8	—	—	—	—	—	—	—
	9	—	—	—	—	—	—	—
	10	—	—	—	—	—	—	—
	11	—	—	—	—	—	—	—
	12	—	0,0	1,0	—	—	—	1,0
	13	—	—	—	—	—	—	—
	14	—	—	—	—	—	2,2	2,2
	15	2,4	3,9	4,8	0,0	0,1	0,2	11,4
	16	0,3	0,3	0,0	0,0	0,0	0,0	0,6
	17	—	—	—	—	—	—	—
	18	—	—	—	—	—	—	—
	19	—	—	—	—	0,0	—	0,0
	20	—	—	—	—	0,0	0,0	0,0
	21	0,5	0,3	9,6	0,0	—	—	10,4
	22	—	—	—	—	—	—	—
	23	—	—	—	—	—	—	—
	24	—	—	0,0	—	—	—	0,0
	25	—	—	—	—	—	—	—
	26	—	—	—	—	—	—	—
	27	—	—	—	—	—	0,0	0,0
	28	0,0	4,2	8,4	0,1	0,0	0,0	12,7
	29	0,0	0,0	—	—	—	—	0,0
	30	—	—	—	—	—	—	—
	31	—	0,0	—	—	0,0	—	0,0
	Sum	5,8	17,6	25,5	0,3	0,1	2,4	51,6

TABLE VI^{a}. AMOUNT OF PRECIPITATION.

Month & Year	Day	2h am	6h am	10h am	2h pm	6h pm	10h pm	Sum
YAMAGUCHI 35m.								
		mm	mm	mm	mm	mm	mm	mm
IX 1889	28	—	—	0,5	—	—	0,0	0,5
	29	—	—	0,0	—	—	—	0,0
	30	—	—	—	—	—	—	—
	Sum	7,3	32,7	78,9	18,1	4,7	13,7	155,4
X 1889	1	—	—	—	—	—	—	—
	2	—	—	—	—	—	—	—
	3	—	—	—	—	—	—	—
	4	—	—	—	—	—	—	—
	5	—	—	—	—	—	—	—
	6	3,0	13,7	3,5	0,3	—	—	20,5
	7	—	—	—	0,0	—	—	0,0
	8	—	—	—	—	—	—	—
	9	—	—	—	—	—	—	—
	10	—	—	—	—	—	—	—
	11	—	—	—	—	—	—	—
	12	—	—	1.7		—	—	1.7
	13	—	—	—	—	—	—	—
	14	—	—	—	—	—	3,7	3,7
	15	5,8	1,6	25,6	0,2	0,1	0,1	33,4
	16	—	—	—	0,0	—	—	0,0
	17	—	—	—	—	—	—	—
	18	—	—	—	—	—	—	—
	19	—	—	—	—	—	—	—
	20	—	—	—	—	0,0	0,0	0,0
	21	0,0	0,0	2,5	0,0	—	—	2,5
	22	—	—	—	—	—	—	—
	23	—	—	—	—	—	—	—
	24	—	—	0,0	—	—	—	0,0
	25	—	—	—	—	—	—	—
	26	—	—	—	—	—	—	—
	27	—	—	—	—	—	—	—
	28	2,6	5,8	14,2	1,2	0,0	—	23,8
	29	—	—	—	—	—	—	—
	30	—	—	—	—	—	—	—
	31	—	—	—	—	—	—	—
	Sum	11,4	21,1	47,5	1,7	0,1	3,8	85,6

TABLE VII. AMOUNT OF CLOUDS.

Month & Year	Day	2h am	4h am	6h am	7h am	8h am	9h am	10h am	11h am	Noon	1h pm	2h pm	3h pm	4h pm	5h pm	6h pm	8h pm	10h pm	Mid-night	Mean
									FUJI 3718m.											
		0–10	0–10	0–10	0–10	0–10	0–10	0–10	0–10	0–10	0–10	0–10	0–10	0–10	0–10	0–10	0–10	0–10	0–10	0–10
VIII 1889	1	8	9	9	...	10	...	7	...	9	...	9	...	8	...	7	6	0	0	6,8
	2	0	1	6	...	3	...	3	...	3	...	4	...	9	...	8	1	0	0	2,9
	3	1	5	3	...	4	...	3	...	3	...	4	...	2	...	8	0	0	0	2,7
	4	0	3	1	...	1	...	0	...	9	...	3	...	3	...	9	7	4	2	3,5
	5	0	3	2	...	1	...	1	...	3	...	2	...	2	...	3	1	0	0	1,5
	6	0	0	0	...	0	...	0	...	1	...	0	...	1	...	1	0	0	0	0,2
	7	0	0	0	...	0	...	1	...	1	...	1	...	6	...	1	3	1	5	1,6
	8	0	1	2	...	8	...	10	10	9	5	5	8	9	10	7	10	10	10	6,8
	9	10	10	10	3	1	1	1	2	1	3	8	2	9	9	9	9	10	4	6,8
	10	1	0	1	1	2	2	1	1	4	1	3	1	4	4	1	8	10	7	3,6
	11	10	10	8	7	6	4	0	1	0	0	1	1	4	3	0	1	6	0	3,6
	12	1	2	0	0	0	1	1	1	1	1	9	5	9	3	3	8	8	1	3,6
	13	10	10	10	10	10	10	7	5	1	3	9	10	10	8	9	0	3	1	6,7
	14	10	10	0	0	0	3	9	8	8	10	7	9	10	10	10	10	10	10	7,8
	15	0	0	0	0	0	0	0	0	0	1	2	0	0	0	0	0	0	0	0,2
	16	0	0	0	0	0	0	1	0	1	1	1	2	1	1	1	0	0	1	0,5
	17	1	1	1	0	1	0	0	1	2	1	1	10	10	10	10	10	10	10	4,7
	18	10	10	10	10	10	10	10	10	10	10	10	10	10	10	10	10	10	10	10,0
	19	10	10	10	10	10	10	10	10	10	10	10	10	10	10	10	10	10	10	10,0
	20	10	10	10	10	10	10	10	10	10	10	10	10	10	10	10	10	10	10	10,0
	21	10	10	10	10	10	10	5	10	0	0	0	0	0	0	0	0	0	0	3,7
	22	0	0	0	0	0	0	0	0	0	0	0	10	0	0	0	0	0	0	0,6
	23	0	0	1	4	1	1	0	0	0	6	8	10	10	10	10	10	3	1	3,7
	24	0	0	0	0	1	6	7	5	4	10	8	5	9	6	8	0	0	0	3,1
	25	0	0	0	0	1	1	2	4	1	1	1	6	2	10	8	0	8	0	1,9
	26	0	0	7	8	9	9	9	7	2	10	10	10	10	10	10	10	10	10	7,2
	27	10	2	7	6	10	10	10	10	10	10	10	10	10	10	7	3	1	10	7,5
	28	10	0	10	10	10	10	10	10	10	1	10	10	10	10	10	10	10	10	9,2
	29	10	10	10	10	10	10	10	10	10	10	10	10	10	10	10	10	0	10	9,2
	30	10	8	9	10	0	0	1	0	2	0	0	0	7	10	9	1	0	0	3,9
	31	0	0	0	0	0	1	8	10	10	9	8	10	10	10	10	10	10	10	6,3
	Mean	4,3	4,0	4,4	4,7	4,2	4,7	4,4	5,2	4,4	4,3	5,3	6,6	6,4	7,1	6,4	5,1	4,6	4,3	4,8
IX 1889	1	8	..	10	...	...	...	10	...	...	...	10	...	...	...	10	...	10	...	9,7
	2	10	..	10	...	...	...	10	...	...	...	10	...	...	...	10	...	10	...	10,0
	3	10	...	10	...	...	...	10	...	...	...	1	...	...	...	7	...	0	...	6,3
	4	0	...	10	...	...	...	10	...	...	...	10	...	...	...	10	...	10	...	8,3
	5	10	...	10	...	...	...	10	...	...	...	10	...	...	...	10	...	10	...	10,0
	6	10	...	10	...	...	...	10	...	...	...	10	...	...	...	1	...	4	...	7,5
	7	6	...	1	...	...	...	3	...	...	...	6	...	...	...	10	...	10	...	6,0
General Mean		4,9	4,0	5,2	4,7	4,2	4,7	5,2	5,2	4,4	4,3	5,8	6,6	6,4	7,1	6,7	5,1	5,2	4,3	5,4
									ONTAKE 3062m.											
VIII 1891	1	0	0	0	...	0	...	0	...	1	...	2	...	1	...	6	0	10	10	2,5
	2	10	10	10	...	10	...	10	...	10	...	10	...	10	...	10	10	10	10	10,0
	3	10	10	10	...	10	...	10	...	10	...	10	...	10	...	10	10	10	10	10,0
	4	10	10	10	...	10	...	10	...	10	...	10	...	10	...	10	1	0	10	8,4
	5	10	10	9	...	10	...	10	...	8	...	10	...	10	...	10	10	10	10	9,7
	6	10	10	10	...	8	...	10	...	10	...	10	...	10	...	9	10	0	0	8,1
	7	10	10	8	...	9	...	7	...	5	...	1	...	0	...	10	1	1	1	5,2
	8	0	0	0	...	3	...	8	...	7	...	8	...	6	...	7	0	0	0	3,2
	9	0	0	3	...	6	...	9	...	10	...	10	...	10	...	0	0	0	10	4,8
	10	10	10	10	...	10	...	10	...	10	...	10	...	10	...	10	10	10	10	10,0
	11	10	10	10	...	10	...	10	...	10	...	10	...	10	...	10	9	10	10	9,9
	12	0	0	5	...	1	...	1	...	0	...	9	...	8	...	5	1	10	10	4,2
	13	10	10	10	...	10	...	3	...	6	...	10	...	10	...	8	7	4	0	7,3
	14	0	2	2	...	0	...	6	...	9	...	9	...	10	...	2	6	0	0	3,8
	15	0	0	0	...	0	...	0	...	0	...	10	...	10	...	0	0	10	0	2,5
	16	0	10	0	...	10	...	10	...	10	...	10	...	10	...	10	10	10	10	8,3
	17	10	10	10	...	10	...	10	...	10	...	10	...	10	...	10	10	10	10	10,0
	18	10	10	10	...	10	...	10	...	10	...	10	...	10	...	9	2	10	10	9,2

Month & Year	Day	2h am	4h am	6h am	7h am	8h am	9h am	10h am	11h am	Noon	1h pm	2h pm	3h pm	4h pm	5h pm	6h pm	7h pm	8h pm	9h pm	10h pm	11h pm	Mid-night	Mean
YAMANAKA 990m.																							
VIII 1889		0–10	0–10	0–10	0–10	0–10	0–10	0–10	0–10	0–10	0–10	0–10	0–10	0–10	0–10	0–10	0–10	0–10	0–10	0–10	0–10	0–10	0–10
	1	3	..	3	3	3	3	4	4	5	2	2	5	5	3	2	1	0	0	0	..	..	2,3
	2	1	..	2	1	1	2	4	3	2	3	2	3	3	5	4	3	1	1	0	..	..	2,2
	3	0	..	5	3	3	3	2	2	2	2	3	2	3	1	5	6	1	0	0	..	..	2,5
	4	5	..	1	1	2	3	5	3	5	9	9	9	6	6	7	5	9	6	3	..	..	5,0
	5	0	..	2	2	5	7	3	5	4	4	3	3	6	4	3	3	3	0	1	..	..	2,3
	6	0	..	0	0	0	1	1	1	3	3	3	2	4	2	3	8	10	10	10	..	..	2,8
	7	0	..	3	1	1	1	2	2	4	3	2	1	3	2	3	1	4	3	9	..	..	3,2
	8	5	..	5	3	4	6	9	6	7	8	9	4	7	9	5	6	9	9	10	..	..	7,2
	9	10	..	10	10	10	6	5	6	7	6	4	3	6	6	8	6	6	8	10	10	10	7,8
	10	4	..	1	1	4	4	9	6	5	4	3	4	6	8	10	10	6	10	10	10	10	6,2
	11	10	..	10	6	6	9	5	7	4	5	5	8	7	8	10	10	10	10	10	7	4	8,3
	12	10	..	10	9	8	9	6	5	7	4	6	5	10	8	9	5	4	10	9	10	10	8,3
	13	10	..	8	5	2	2	4	4	4	5	10	9	10	9	10	9	2	3	8	10	9	8,3
	14	6	..	0	0	2	2	8	8	6	4	7	7	9	6	10	10	10	10	10	1	1	6,8
	15	0	..	0	10	2	6	4	5	3	5	5	2	0	0	0	1	0	0	0	0	0	1,5
	16	0	..	5	1	1	2	1	3	3	8	9	7	7	9	0	1	10	10	1	0	0	2,7
	17	7	..	1	0	0	0	1	1	6	5	4	6	9	10	10	10	9	10	10	..	..	5,5
	18	10	..	10	10	10	10	10	10	10	10	10	10	10	10	10	10	10	10	10	10	10	10,0
	19	10	..	10	10	10	10	10	10	10	10	10	10	10	10	10	10	10	9	10	..	10	10,0
	20	10	..	10	10	10	10	10	10	10	10	10	10	10	10	10	10	10	10	10	10	10	10,0
	21	10	..	10	10	8	10	7	7	4	3	1	8	5	2	3	1	2	4	0	..	0	5,2
	22	10	..	10	1	2	1	1	7	6	1	3	8	9	4	0	0	10	0	0	..	0	4,0
	23	0	..	2	4	2	1	1	1	2	2	3	9	10	10	10	10	10	10	10	10	0	4,3
	24	0	..	0	0	0	0	1	2	5	6	9	8	9	9	9	8	0	0	0	10	9	3,2
	25	9	..	8	1	6	5	8	9	9	6	9	9	10	10	10	10	10	3	7	0	10	8,5
	26	1	..	5	8	9	9	9	8	4	7	9	10	10	10	10	10	10	10	10	10	10	7,3
	27	10	..	10	10	10	10	10	10	10	10	10	10	10	10	9	10	10	10	10	10	10	9,8
	28	10	..	10	10	10	10	10	10	10	10	10	10	10	10	10	10	10	10	10	10	10	10,0
	29	10	..	10	10	10	10	10	10	10	10	10	10	10	10	10	10	10	10	10	10	10	10,0
	30	10	..	10	10	10	10	10	10	10	8	10	10	10	10	10	10	3	10	10	10	10	10,0
	31	10	..	10	10	10	8	6	10	10	10	8	10	10	10	10	10	10	10	10	10	10	9,0
	Mean	5,9	..	5,8	5,2	5,2	5,5	5,7	6,0	6,0	5,9	6,4	6,8	7,5	7,1	7,1	6,9	6,7	6,6	6,7	4,8	4,9	6,3
IX 1889	1	9	..	9				10				10				10				8			9,3
	2	10	..	10				10				9				10				10			9,8
	3	10	..	10				10				10				10				4			9,0
	4	7	..	10				10				10				10				10			9,5
	5	10	..	10				10				10				10				10			10,0
	6	0	..	9				6				10				7				9			6,8
	7	9	..	10				10				10				10				10			9,8
General Mean		6,8	..	6,5	5,2	5,2	5,5	6,4	6,0	6,0	5,9	7,0	6,8	7,5	7,1	7,5	6,9	6,7	6,6	7,1	4,8	4,9	6,8
KUROSAWA 832m.																							
VIII 1891	1	0	0	5	..	0	..	0	..	6	..	7	..	5	..	8	..	0	..	2	..	7	3,3
	2	5	10	10	..	10	..	10	..	10	..	10	..	10	..	10	..	10	..	4	..	8	8,9
	3	10	10	10	..	10	..	10	..	9	..	5	..	5	..	8	..	10	..	3	..	9	8,2
	4	10	10	10	..	10	..	10	..	10	..	10	..	10	..	10	..	5	..	10	..	10	9,6
	5	10	10	10	..	10	..	10	..	10	..	4	..	10	..	10	..	10	..	10	..	10	9,5
	6	10	10	10	..	9	..	6	..	10	..	10	..	9	..	10	..	3	..	10	..	9	8,8
	7	10	10	10	..	10	..	8	..	10	..	10	..	10	..	10	..	6	..	0	..	5	8,2
	8	1	2	4	..	3	..	8	..	6	..	7	..	6	..	0	..	1	..	0	..	5	3,6
	9	8	6	9	..	7	..	9	..	7	..	4	..	2	..	0	..	0	..	0	..	0	4,3
	10	5	9	9	..	9	..	10	..	9	..	10	..	10	..	10	..	6	..	9	..	8	8,7
	11	2	3	7	..	9	..	10	..	10	..	10	..	8	..	10	..	10	..	6	..	4	7,4
	12	1	0	5	..	2	..	3	..	3	..	5	..	10	..	3	..	2	..	0	..	8	3,5
	13	10	1	10	..	8	..	6	..	7	..	9	..	7	..	10	..	9	..	3	..	0	6,7
	14	0	0	0	..	0	..	2	..	3	..	6	..	7	..	7	..	10	..	9	..	0	3,7
	15	10	9	9	..	4	..	0	..	5	..	7	..	0	..	0	..	0	..	1	..	0	3,7
	16	0	0	0	..	0	..	2	..	1	..	10	..	10	..	10	..	10	..	10	..	10	5,2
	17	10	10	10	..	10	..	10	..	10	..	10	..	10	..	10	..	10	..	10	..	10	10,0
	18	10	10	10	..	10	..	10	..	10	..	10	..	10	..	10	..	1	..	8	..	2	8,4

TABLE VII. AMOUNT OF CLOUDS.

Month & Year	Day	2h am	4h am	6h am	7h am	8h am	9h am	10h am	11h am	Noon	1h pm	2h pm	3h pm	4h pm	5h pm	6h pm	8h pm	10h pm	Mid-night	Mean
		ONTAKE 3062m.																		
VIII		0–10	0–10	0–10	0–10	0–10	0–10	0–10	0–10	0–10	0–10	0–10	0–10	0–10	0–10	0–10	0–10	0–10	0–10	0–10
1894	19	10	0	0	...	0	...	0	...	0	...	0	...	1	...	1	0	4	2	1,5
	20	2	1	1	...	2	...	10	...	10	...	5	...	8	...	8	9	10	10	6,3
	21	10	10	10	...	10	...	10	...	10	...	10	...	10	...	10	10	10	10	10,0
	22	10	10	10	...	10	...	10	...	10	...	10	...	10	...	10	10	10	10	10,0
	23	10	10	10	...	10	...	10	...	10	...	10	...	10	...	10	1	9	1	8,4
	24	4	10	9	...	10	...	10	...	10	...	10	...	10	...	10	10	10	9	9,3
	25	10	10	10	...	8	...	10	...	10	...	8	...	6	...	2	0	0	0	6,2
	26	0	0	0	...	0	...	1	...	10	...	10	...	1	...	0	0	0	0	1,8
	27	0	0	0	...	4	...	4	...	10	...	10	...	10	...	1	0	0	6	4,0
	28	9	3	10	...	10	...	10	...	10	...	10	...	8	...	3	0	0	0	6,1
	29	10	10	10	...	10	...	10	...	10	...	10	...	6	...	8	0	0	0	7,0
	30	0	0	10	...	8	...	3	...	6	...	3	...	8	...	0	0	0	0	3,2
	31	0	0	1	...	0	...	1	...	10	...	10	...	10	...	0	0	0	0	2,7
	Mean	6,0	6,0	6,4		6,7		7,2		8,1		8,5		8,2		6,5	4,4	5,4	5,4	6,6
IX	1	0	0	1	...	0	...	9	...	9	...	10	...	2	...	10	0	0	10	4,2
1894	2	10	9	7	...	10	...	10	...	10	...	10	...	10	...	10	0	0	0	7,2
	3	0	0	3	...	3	...	5	...	10	...	10	...	10	...	7	0	0	0	4,0
	4	0	0	0	...	1	...	0	...	10	...	10	...	9	...	0	0	0	0	2,5
	5	0	0	0	...	0	...	0	...	0	...	0	...	8	...	0	0	0	0	0,7
	6	0	0	0	...	0	...	0	...	10	...	10	...	7	...	0	0	0	0	2,2
	7	0	2	9	...	7	...	9	...	10	...	10	...	10	...	10	8	9	10	7,8
	8	10	10	10	...	10	...	10	...	10	...	10	...	10	...	10	10	10	10	10,0
	9	10	10	10	...	10	...	10	...	10	...	10	...	10	...	10	10	10	10	10,0
	10	10	10	10	...	10	...	10	...	10	...	10	...	10	...	10	10	10	10	10,0
	11	10	10	10	...	10	...	10	...	10	...	10	...	10	...	10	0	10	10	9,2
	12	10	4	0	...	0	...	10	...	10	...	8	...	8	...	0	10	10	10	6,7
General Mean		5,7	5,6	6,0		6,3		7,1		8,4		8,7		8,3		6,5	4,3	5,2	5,6	6,5
		GOZAISHODAKE 1200m.																		
IX	4	10	10	10	...	10	...	10	...	10	...	10	...	10	...	10	10	10	10	10,0
1888	5	10	10	8	...	10	...	10	...	10	...	7	...	8	...	7	6	1	0	7,2
	6	0	0	1	...	1	...	0	...	0	...	1	...	1	...	4	2	2	10	1,9
	7	10	1	7	...	10	...	8	...	6	...	5	...	4	...	5	1	1	9	4,7
	8	7	10	10	...	10	...	10	...	10	...	10	...	10	...	10	10	10	10	9,8
	9	10	9	7	...	8	...	10	...	10	...	10	...	10	...	10	9	9	2	9,3
	10	10	3	10	...	10	...	10	...	10	...	10	...	10	...	10	10	10	10	8,7
	11	10	10	10	...	10	...	10	...	10	...	10	...	10	...	10	10	10	10	10,0
	12	10	10	10	...	10	...	10	...	10	...	10	...	10	...	10	10	10	10	10,0
	13	10	0	0	...	0	...	2	...	2	...	10	...	2	...	10	10	10	6	5,1
	14	10	10	4	...	8	...	1	...	9	...	6	...	9	...	8	10	10	4	7,4
	15	5	10	7	...	7	...	3	...	3	...	6	...	9	...	10	10	10	10	7,5
	16	10	10	10	...	10	...	10	...	10	...	10	...	10	...	5	7	5	10	8,9
	17	10	10	10	...	8	...	5	...	2	...	1	...	1	...	2	0	0	0	4,1
	18	0	0	10	...	2	...	1	...	1	...	1	...	0	...	1	1	0	1	1,5
	19	0	0	1	...	10	...	10	...	10	...	10	...	10	...	10	10	10	10	7,6
	20	10	10	10	...	10	...	10	...	10	...	10	...	10	...	10	10	10	10	10,0
	21	10	10	10	...	10	...	10	...	10	...	10	...	10	...	10	10	10	10	10,0
	22	10	10	10	...	10	...	10	...	10	...	10	...	10	...	10	10	10	10	10,0
	23	10	10	10	...	10	...	9	...	5	...	1	...	1	...	1	1	1	3	5,1
	24	1	1	4	...	9	...	3	...	8	...	9	...	10	...	10	8	7	9	6,6
	25	9	10	9	...	10	...	10	...	10	...	10	...	10	...	10	10	10	10	9,9
	26	10	10	9	...	3	...	10	...	10	...	10	...	10	...	1	0	1	4	6,5
	27	10	1	1	...	1	...	7	...	8	...	7	...	9	...	10	9	0	1	5,3
	28	0	0	0	...	1	...	0	...	0	...	1	...	4	...	9	3	1	0	1,6
	29	0	0	0	...	1	...	2	...	8	...	9	...	5	...	10	3	1	1	3,3
	30	1	7	4	...	5	...	5	...	5	...	8	...	10	...	10	10	10	10	7,0
X	1	10	10	10	...	10	...	10	...	6	...	10	...	8	...	8	7	3	7	9,0
1888	2	8	7	1	...	0	...	1	...	1	...	1	...	1	...	0	0	0	0	1,7
	3	0	0	0	...	0	...	0	...	0	...	0	...	1	...	0	0	0		0,0
	Mean	7,0	6,3	6,4	...	6,8	...	6,6	...	6,8	...	7,1	...	7,1	...	7,4	6,6	5,7	6,3	6,3

Month & Year	Day	2h am	4h am	6h am	7h am	8h am	9h am	10h am	11h am	Noon	1h pm	2h pm	3h pm	4h pm	5h pm	6h pm	7h pm	8h pm	9h pm	10h pm	11h pm	Mid-night	Mean
KUROSAWA 832m.																							
VIII 1891		0–10	0–10	0–10	0–10	0–10	0–10	0–10	0–10	0–10	0–10	0–10	0–10	0–10	0–10	0–10	0–10	0–10	0–10	0–10	0–10	0–10	0–10
	19	9	4	3	..	0	..	1	..	7	..	5	..	3	..	6	..	0	..	5	..	0	3.6
	20	7	1	10	..	3	..	7	..	9	..	6	..	7	..	7	..	8	..	10	..	10	7.1
	21	10	10	10	..	10	..	10	..	10	..	10	..	10	..	10	..	10	..	10	..	10	10.0
	22	10	10	10	..	10	..	10	..	10	..	10	..	10	..	10	..	10	..	10	..	10	10.0
	23	10	10	10	..	10	..	10	..	10	..	10	..	10	..	10	..	9	..	9	..	2	9.2
	24	10	10	10	..	9	..	10	..	10	..	10	..	10	..	10	..	10	..	10	..	10	9.9
	25	9	10	10	..	4	..	9	..	10	..	7	..	7	..	1	..	0	..	7	..	7	6.7
	26	3	5	10	..	10	..	0	..	0	..	3	..	1	..	0	..	0	..	0	..	0	2.7
	27	0	0	0	..	3	..	0	..	4	..	8	..	8	..	7	..	9	..	0	..	5	3.7
	28	9	6	1	..	9	..	8	..	10	..	3	..	1	..	1	..	0	..	0	..	0	4.0
	29	3	9	10	..	9	..	9	..	4	..	2	..	0	..	9	..	0	..	0	..	0	4.6
	30	0	1	10	..	9	..	1	..	1	..	2	..	2	..	0	..	0	..	0	..	0	2.3
	31	0	0	1	..	0	..	0	..	0	..	2	..	1	..	0	..	0	..	1	..	10	1.2
	Mean	6.2	6.0	7.5	..	6.7	..	6.4	..	7.1	..	7.2	..	6.7	..	6.7	..	5.1	..	5.1	..	5.5	6.3
IX 1891	1	1	2	9	..	0	..	3	..	2	..	6	..	1	..	4	..	0	..	3	..	0	2.6
	2	1	1	10	..	1	..	9	..	9	..	9	..	9	..	5	..	0	..	3	..	1	4.8
	3	1	8	10	..	6	..	8	..	8	..	9	..	6	..	5	..	2	..	2	..	10	6.2
	4	2	0	9	..	10	..	8	..	1	..	1	..	2	..	0	..	0	..	0	..	0	2.7
	5	2	10	10	..	0	..	0	..	0	..	0	..	6	..	0	..	0	..	0	..	0	2.3
	6	1	10	10	..	0	..	0	..	1	..	1	..	1	..	0	..	0	..	1	..	0	2.1
	7	0	0	10	..	9	..	8	..	10	..	10	..	10	..	10	..	8	..	10	..	10	7.9
	8	10	10	9	..	10	..	10	..	10	..	10	..	10	..	10	..	10	..	10	..	10	9.9
	9	10	10	10	..	10	..	10	..	10	..	10	..	10	..	10	..	10	..	9	..	10	9.9
	10	10	10	10	..	10	..	8	..	9	..	7	..	2	..	10	..	6	..	10	..	1	7.7
	11	1	0	9	..	1	..	2	..	1	..	1	..	1	..	0	..	0	..	0	..	0	1.3
	12	6	0	0	..	0	..	0	..	0	..	0	..	1	..	1	..	0	..	10	..	10	2.3
General Mean		5.5	5.7	7.9	..	6.1	..	6.2	..	6.6	..	6.6	..	6.2	..	6.1	..	4.5	..	5.0	..	5.1	6.0
YOKKAICHI 4m.																							
IX 1888	4	9		9				9				10				9				5			8.5
	5	10		8				10				4				7				0			6.5
	6	0		0				0				0				1				0			0.2
	7	1		9				3				1				2				0			2.7
	8	0		7				10				10				10				10			7.8
	9	10		10				10				10				10				7			9.5
	10	10		10				10				10				10				10			10.0
	11	10		10				10				10				7				8			9.2
	12	10		10				9				10				10				10			9.8
	13	0		0				0				9				1				1			1.8
	14	0		1				0				1				9				10			3.5
	15	5		8				6				5				10				10			7.3
	16	10		10				10				10				9				2			8.5
	17	3		0				0				0				2				0			0.8
	18	0		6				1				0				1				0			1.3
	19	0		9				9				9				10				10			7.8
	20	10		10				10				10				10				10			10.0
	21	10		10				10				10				10				10			10.0
	22	10		10				10				10				10				3			8.8
	23	0		0				1				0				0				0			0.2
	24	0		7				9				9				10				9			7.3
	25	9		9				10				10				10				10			9.7
	26	10		10				3				1				0				10			5.7
	27	10		2				6				9				10				0			6.2
	28	0		1				0				1				10				0			2.0
	29	0		1				2				3				5				0			1.8
	30	0		6				7				8				10				10			6.8
X 1888	1	10		10				10				10				9				5			9.0
	2	0		0				0				0				0				0			0.0
	3	0		0				0				0				0				0			0.0
	Mean	4.9	..	6.1	..	..	..	5.8	..	..	..	6.0	..	..	..	6.7	..	..	..	5.0	..	..	5.8

TABLE VII. AMOUNT OF CLOUDS.

Month & Year	Day	2h am	6h am	10h am	2h pm	6h pm	10h pm	Mean
HIGASHI HOREN 736m.								
VIII 1889		0–10	0–10	0–10	0–10	0–10	0–10	0–10
	1	9	10	9	3	9	0	6,7
	2	10	10	8	10	6	2	7,7
	3	0	1	2	4	6	0	2,2
	4	0	9	7	9	4	3	5,3
	5	0	1	1	2	1	0	0,8
	6	0	10	9	4	2	10	5,8
	7	10	10	7	3	1	0	5,2
	8	9	10	7	6	1	0	5,5
	9	0	0	9	8	9	3	4,8
	10	0	5	10	9	9	0	5,5
	11	10	7	4	7	4	0	5,3
	12	0	4	4	8	0	0	2,7
	13	10	7	9	5	10	0	6,8
	14	10	10	10	7	10	10	9,5
	15	10	10	10	10	10	10	10,0
	16	10	10	10	8	10	10	9,7
	17	10	10	10	9	1	0	6,7
	18	0	0	5	9	9	2	4,2
	19	9	10	10	10	10	10	9,8
	20	10	10	10	9	10	9	9,8
	21	2	10	6	10	2	0	5,0
	22	0	10	10	5	10	0	5,8
	23	10	10	9	10	10	10	9,8
	24	10	10	10	10	10	10	10,0
	25	10	10	10	10	10	10	10,0
	26	10	10	10	10	10	10	10,0
	27	10	10	5	2	9	1	6,2
	28	10	9	9	8	10	10	9,3
	29	0	9	8	10	4	5	6,0
	30	10	10	9	10	1	2	7,0
	31	10	10	10	10	10	8	9,7
	Mean	6,42	8,13	7,97	7,58	6,71	4,35	6,86
IX 1889	1	10	10	10	10	10	10	10,0
	2	10	10	10	10	10	10	10,0
	3	0	4	1	1	10	8	4,0
	4	8	10	7	9	10	9	8,8
	5	10	10	10	10	10	10	10,0
	6	10	10	10	10	6	1	7,8
	7	1	8	4	1	8	0	3,7
	8	8	5	1	3	10	2	4,8
	9	3	5	3	0	1	10	3,7
	10	1	6	0	10	10	10	6,2
	11	10	10	10	10	10	10	10,0
	12	10	6	2	10	5	6	6,5
	13	7	0	0	1	2	0	1,7
	14	0	0	0	1	1	2	0,7
	15	4	2	2	8	1	0	2,8
	16	0	0	1	1	0	0	0,3
	17	1	9	4	8	10	1	5,5
	18	0	1	6	8	10	2	4,5
	19	8	10	9	10	10	10	9,5
	20	10	10	10	10	10	10	10,0
	21	10	10	10	10	5	9	9,0
	22	5	10	9	10	7	10	8,5
	23	10	8	10	10	10	0	8,0
	24	1	10	9	7	3	0	5,0
	25	10	10	10	10	10	2	8,7
	26	0	1	5	5	10	10	5,2
	27	10	10	10	10	10	10	10,0

TABLE VIIa. AMOUNT OF CLOUDS.

Month & Year	Day	2h am	6h am	10h am	2h pm	6h pm	10h pm	Mean
YAMAGUCHI 35m.								
VIII 1889		0–10	0–10	0–10	0–10	0–10	0–10	0–10
	1	3	3	6	3	2	1	3,0
	2	2	9	3	4	6	1	4,2
	3	0	10	2	3	6	0	3,5
	4	0	9	5	8	4	5	5,2
	5	0	1	2	2	1	1	1,2
	6	2	10	3	3	1	8	4,5
	7	9	9	3	1	1	0	3,8
	8	1	9	4	2	0	0	2,7
	9	0	2	3	4	5	5	3,2
	10	1	3	7	7	7	0	4,2
	11	9	7	3	6	3	1	4,8
	12	1	5	6	4	1	0	2,8
	13	0	0	5	4	9	5	3,8
	14	9	9	9	5	9	8	8,2
	15	10	9	9	10	10	9	9,5
	16	7	9	9	6	9	3	7,2
	17	9	8	7	7	1	0	5,3
	18	1	1	4	9	8	3	4,3
	19	6	7	8	10	10	10	8,5
	20	10	10	9	6	3	0	6,3
	21	1	7	4	6	3	1	3,7
	22	1	10	7	4	6	0	4,7
	23	8	8	8	6	10	10	8,3
	24	10	10	10	9	8	9	9,3
	25	10	10	10	10	10	10	10,0
	26	10	5	5	4	2	1	4,5
	27	1	2	3	2	8	3	3,2
	28	5	8	7	5	9	10	7,3
	29	0	0	7	9	5	9	6,0
	30	10	10	8	9	3	2	7,0
	31	9	10	10	10	10	8	9,5
	Mean	4,68	6,77	6,00	5,74	5,48	3,97	5,44
IX 1889	1	10	10	9	10	10	10	9,8
	2	10	10	10	10	9	5	9,0
	3	0	3	1	2	8	3	2,8
	4	3	5	8	9	8	9	7,0
	5	10	10	10	6	8	10	9,0
	6	10	10	7	6	2	0	5,8
	7	1	8	6	2	0	0	2,8
	8	0	0	0	8	1	2	1,8
	9	9	8	1	1	0	0	3,2
	10	0	0	1	10	10	10	5,2
	11	10	10	9	10	10	5	9,0
	12	3	2	3	10	5	7	5,0
	13	0	1	1	2	0	0	0,7
	14	0	0	0	2	0	1	0,5
	15	4	6	3	9	4	0	3,7
	16	0	0	1	3	0	0	0,7
	17	1	9	1	6	10	0	4,5
	18	1	0	2	4	9	1	2,8
	19	9	9	8	9	10	10	9,2
	20	10	10	10	10	10	10	10,0
	21	10	10	10	10	7	3	8,3
	22	3	7	9	10	9	8	7,7
	23	10	10	9	10	6	0	7,5
	24	0	6	8	7	5	0	4,3
	25	1	10	10	10	10	0	6,8
	26	0	1	3	7	10	10	5,2
	27	10	10	10	10	10	10	10,0

TABLE VII. AMOUNT OF CLOUDS.

Month & Year	Day	2h am	6h am	10h am	2h pm	6h pm	10h pm	Mean
HIGASHI HOBEN 736m.								
		0–10	0–10	0–10	0–10	0–10	0–10	0–10
IX 1889	28	10	5	10	10	10	10	9,2
	29	10	10	3	5	8	10	7,7
	30	9	5	7	5	1	2	4,8
	Mean	6,29	6,63	6,10	7,10	7,27	5,80	6,52
X 1889	1	0	8	5	7	9	6	5,8
	2	9	10	8	10	9	8	9,0
	3	10	6	3	0	1	10	5,0
	4	2	1	2	4	1	2	2,0
	5	10	8	1	4	9	10	7,0
	6	10	10	10	10	10	10	10,0
	7	10	0	2	2	9	10	5,5
	8	10	1	2	1	8	10	5,3
	9	10	10	10	8	9	10	9,5
	10	10	1	4	3	8	5	5,2
	11	9	9	4	5	8	10	7,5
	12	10	10	8	2	5	7	7,0
	13	8	0	1	2	0	1	2,0
	14	8	4	3	10	10	10	7,5
	15	10	10	10	10	10	10	10,0
	16	10	10	1	8	10	0	6,5
	17	0	1	0	10	2	10	3,8
	18	10	1	1	10	4	0	4,3
	19	6	10	9	10	9	0	7,3
	20	1	1	7	10	10	10	6,5
	21	10	10	10	10	6	0	7,7
	22	1	9	3	9	6	2	5,0
	23	0	1	8	7	8	3	4,5
	24	10	10	10	10	9	6	9,2
	25	2	1	1	3	6	1	2,3
	26	1	10	6	1	5	10	5,5
	27	10	10	1	1	4	10	6,0
	28	10	10	10	10	10	10	10,0
	29	10	10	9	6	5	0	6,7
	30	0	8	8	1	0	4	3,5
	31	10	9	7	10	9	8	8,8
	Mean	7,00	6,42	5,29	6,26	6,74	6,32	6,32

Month & Year	Day	2h am	6h am	10h am	2h pm	6h pm	10h pm	Mean
YAMAGUCHI 35m.								
		0–10	0–10	0–10	0–10	0–10	0–10	0–10
IX 1889	28	2	4	7	10	9	10	7,0
	29	9	10	7	7	4	6	7,2
	30	1	4	5	5	2	9	4,3
	Mean	4,57	6,07	5,63	7,17	6,10	4,68	5,69
X 1889	1	0	0	5	7	7	2	3,5
	2	8	8	7	9	10	9	8,5
	3	10	7	2	1	1	5	4,3
	4	1	2	1	2	4	3	2,2
	5	10	10	1	4	9	10	7,3
	6	10	10	10	10	5	10	9,2
	7	6	2	3	4	7	9	5,2
	8	2	2	1	5	9	10	6,0
	9	10	10	9	8	9	9	9,2
	10	4	1	2	3	9	8	4,5
	11	10	10	4	5	7	9	7,5
	12	9	10	9	2	7	9	7,7
	13	1	0	1	3	1	1	1,2
	14	7	4	4	10	10	10	7,5
	15	10	10	10	10	10	10	10,0
	16	9	7	3	4	4	0	4,5
	17	0	0	1	9	3	6	3,2
	18	8	3	1	10	7	0	4,8
	19	8	8	9	10	10	0	7,5
	20	0	1	3	10	10	10	5,7
	21	10	9	10	10	10	0	8,2
	22	3	7	4	10	6	4	5,7
	23	0	4	5	5	3	2	3,2
	24	10	10	10	10	10	9	9,8
	25	3	2	1	1	4	1	2,0
	26	1	8	5	2	9	10	5,8
	27	10	10	1	3	3	10	6,2
	28	10	10	10	10	9	7	9,3
	29	9	10	7	6	7	0	6,5
	30	0	7	7	1	0	9	4,0
	31	10	9	9	10	9	7	9,0
	Mean	6,32	6,16	5,00	6,26	6,74	6,10	6,10

TABLE VIII. EXTREME AIR TEMPERATURE AND RADIATION.

Month & Year	Day	Max. Temp.	Min. Temp.	Mean. Temp.	Range	Max. S. R.	Max. S.R.V.	Min. T.R.
FUJI 3718ᵐ.								
VIII		°	°	°	°	°	°	°
1889	1	18.9	3.5	11.20	14.4	..	51.5	1.9
	2	21.4	4.2	12.80	17.2	..	51.3	3.0
	3	17.0	4.0	10.50	13.0	..	52.3	2.6
	4	18.1	3.0	10.55	15.1	..	55.4	1.7
	5	20.7	4.5	12.60	16.2	..	48.4	3.6
	6	18.4	3.9	11.15	14.5	..	44.9	3.2
	7	16.5	4.7	10.60	11.8	..	43.2	3.8
	8	18.9	4.4	11.65	14.5	..	55.7	3.7
	9	17.5	3.5	10.50	14.0	..	53.1	3.5
	10	17.4	3.0	10.20	14.4	..	49.2	1.8
	11	14.9	3.6	9.25	11.3	..	44.7	1.0
	12	17.0	2.8	9.90	14.2		52.1	..
	13	14.9	2.7	8.80	12.2		53.0	1.3
	14	16.0	1.8	8.90	14.2		56.8	0.8
	15	18.2	99.5	8.85	18.7		45.3	96.0
	16	19.0	3.3	11.15	15.7	..	48.2	2.2
	17	18.0	2.7	10.35	15.3		49.2	97.2
	18	6.9	2.1	4.50	4.8		16.6	2.3
	19	8.0	3.6	5.80	4.4		21.0	3.6
	20	7.4	3.6	5.50	3.8			..
	21	9.6	4.6	7.10	5.0		40.9	2.0
	22	15.2	3.6	9.40	11.6		45.4	2.9
	23	16.8	4.2	10.50	12.6	..	47.6	3.5
	24	16.0	2.8	9.40	13.2	..	49.0	1.1
	25	14.2	3.1	8.65	11.1		44.5	2.1
	26	7.5	2.0	4.75	5.5		40.4	2.1
	27	12.6	3.1	7.85	9.5		32.3	2.8
	28	10.2	5.3	7.75	4.9		25.6	4.8
	29	10.2	5.2	7.50	5.0	..	20.7	4.3
	30	8.4	0.0	4.20	8.4		44.3	98.5
	31	11.4	99.5	5.45	11.9		40.6	97.6
	Mean	14.75	3.15	8.95	11.60		44.11	19.0
IX	1	12.4	99.1	5.75	13.3	..	..	..
1889	2	7.4	2.9	5.15	4.5	..	..	..
	3	8.5	1.4	4.95	7.1	..	..	..
	4	10.2	1.6	5.90	8.6	..	..	..
	5	6.6	96.7	1.65	9.9	..	..	..
	6	7.9	96.6	2.25	11.3	..	..	..
	7	3.6	1.4	7.50	12.2	..	..	..
General Mean		13.53	2.56	8.17	11.23		44.11	19.0
ONTAKE 3062ᵐ.								
VIII	1	18.8	4.4	..	14.4	24.5	..	2.9
1891	2	10.7	6.2	..	4.5	13.0	..	5.9
	3	9.6	5.8	..	3.8	11.5	..	5.8
	4	10.9	5.2	..	5.7	12.1	..	2.9
	5	14.4	7.5	..	6.9	23.8	..	7.3
	6	13.8	7.3	..	6.5	21.9	..	6.4
	7	18.1	6.3	..	11.8	26.2	..	4.7
	8	17.0	6.1	..	10.9	22.4	..	4.1
	9	16.4	7.2	..	9.2	20.8	..	5.8
	10	11.8	6.3	..	5.5	15.7	..	6.2
	11	10.3	4.7	..	5.6	10.7	..	4.2
	12	17.5	5.0	..	12.5	21.0	..	4.2
	13	18.3	7.6	..	10.7	25.2	..	6.7
	14	21.2	7.2	..	14.0	26.2	..	6.0
	15	17.4	5.3	..	12.1	21.3	..	4.8
	16	9.8	4.3	..	5.5	14.4	..	4.3
	17	10.3	8.2	..	2.1	10.6	..	8.2
	18	9.8	2.7	..	7.1	10.8	..	1.5
	19	12.9	0.8	..	12.1	19.0	..	98.2

TABLE VIIIᵇ. EXTREME AIR TEMPERATURE AND RADIATION.

Month & Year	Day	Max. Temp.	Min. Temp.	Mean. Temp.	Range	Max. S. R.	Max. S.R.V.	Min. T.R.
YAMANAKA 990ᵐ.								
VIII		°	°	°	°	°	°	°
1889	1	29.0	16.2	22.6	12.8	..	67.5	..
	2	30.3	16.6	23.4	13.7	..	64.4	13.8
	3	31.1	17.1	24.1	14.0	..	63.2	14.3
	4	29.6	17.9	23.7	11.7	..	67.0	14.2
	5	27.5	15.9	21.7	11.6	..	71.2	13.0
	6	30.0	15.9	22.9	14.1	..	63.2	12.9
	7	30.6	16.8	23.7	13.8	..	64.9	12.9
	8	28.4	17.2	22.8	11.2	..	64.7	14.6
	9	28.6	18.0	23.3	10.6	..	67.2	16.1
	10	28.6	16.9	22.7	11.7	..	64.7	14.6
	11	27.7	19.4	23.5	8.3	..	65.0	18.0
	12	29.2	19.0	24.1	10.2	..	72.8	17.5
	13	28.0	18.8	23.4	9.2	..	60.0	12.9
	14	26.0	15.1	20.5	10.9	..	62.1	12.0
	15	26.7	14.0	20.3	12.7	..	62.1	11.1
	16	27.6	11.8	19.7	15.8	..	65.5	8.6
	17	28.7	14.1	21.5	14.3	..	67.2	11.4
	18	23.3	14.4	18.8	8.9	..	61.4	16.1
	19	21.6	19.6	20.6	2.0	..	42.6	18.7
	20	20.5	18.0	19.2	2.5	..	24.5	17.2
	21	26.2	19.1	22.6	7.1	..	50.0	15.8
	22	27.9	17.5	22.7	10.4	..	61.8	15.8
	23	29.0	17.1	23.0	11.9	..	50.4	15.4
	24	27.6	16.0	21.8	11.6	..	58.4	15.1
	25	25.6	17.7	21.6	7.9	..	61.7	15.1
	26	24.8	16.5	20.6	8.3	..	..	14.0
	27	21.9	18.0	19.9	3.9	..	32.6	16.0
	28	20.7	18.9	19.8	1.6	..	31.3	18.5
	29	16.1	14.8	15.4	1.3	..	22.4	14.0
	30	18.8	13.4	16.1	5.4	..	42.6	13.0
	31	19.9	12.4	16.1	7.5	..	46.5	12.0
	Mean	26.18	16.59	21.36	9.59	..	57.47	14.47
IX	1	19.6	12.5	16.05	7.1	..	..	..
1889	2	21.5	14.0	17.75	7.5	..	..	..
	3	21.7	15.9	18.80	5.8	..	..	..
	4	19.1	12.7	15.90	6.4	..	..	..
	5	19.5	13.7	16.60	5.8	..	..	..
	6	25.1	16.3	20.70	8.8	..	..	..
	7	19.1	15.7	17.40	3.4	..	..	..
General Mean		25.29	16.29	20.67	9.00	..	57.47	14.47
KUROSAWA 832ᵐ.								
VIII	1	28.3	11.2	..	17.1	36.6	..	8.9
1891	2	27.0	16.3	..	10.7	42.2	..	14.3
	3	28.1	16.8	..	11.3	42.1	..	15.9
	4	26.2	18.1	..	8.1	35.7	..	17.1
	5	26.4	17.2	..	9.2	31.7	..	17.0
	6	27.5	18.7	..	8.8	42.6	..	17.9
	7	28.1	18.0	..	10.1	43.5	..	16.7
	8	31.4	15.3	..	16.1	43.7	..	12.6
	9	31.0	15.4	..	15.6	43.9	..	13.1
	10	29.1	16.4	..	12.7	41.8	..	14.3
	11	28.1	17.2	..	10.9	40.7	..	15.4
	12	30.4	16.2	..	14.2	45.2	..	13.6
	13	31.4	17.0	..	14.4	47.3	..	15.0
	14	32.2	14.4	..	17.8	47.5	..	11.8
	15	31.5	17.1	..	14.4	45.8	..	14.9
	16	30.0	14.0	..	16.0	44.0	..	11.4
	17	26.1	19.1	..	7.0	37.3	..	18.0
	18	26.9	15.0	..	11.9	40.5	..	12.7
	19	27.4	12.0	..	15.4	42.5	..	8.7

TABLE VIII. EXTREME AIR TEMPERATURE AND RADIATION.

Month & Year	Day	Max. Temp.	Min. Temp.	Mean Temp.	Range	Max. S. R.	Max. S.R.V.	Min. T.R.
		ONTAKE 3062m.						
		°	°	°	°	°	°	°
VIII 1891	20	13,9	2,2	..	11,7	21,1	..	90,8
	21	8,5	3,1	..	5,4	8,2	..	2,0
	22	10,2	4,9	..	5,3	10,8	..	4,9
	23	9,5	3,8	..	5,7	9,6	..	1,1
	24	12,8	3,6	..	9,2	17,1	..	1,4
	25	12,8	2,8	..	11,0	21,8	..	97,0
	26	17,0	2,3	..	14,7	24,3	..	95,2
	27	17,8	2,8	..	15,0	27,5	..	97,9
	28	13,7	2,7	..	11,0	20,1	..	0,0
	29	13,8	3,3	..	10,5	21,0	..	98,4
	30	19,6	4,6	..	15,0	28,7	..	0,0
	31	19,8	4,6	..	15,2	29,9	..	1,9
	Mean	14,17	4,80	..	9,37	19,10	..	2,80
IX 1891	1	21,9	5,8	..	16,6	28,0	..	1,7
	2	18,6	6,8	..	11,8	25,6	..	5,9
	3	20,4	8,0	..	12,4	26,0	..	2,8
	4	20,4	6,8	..	13,6	26,3	..	3,5
	5	20,1	5,3	..	14,8	26,3	..	1,2
	6	22,0	6,4	..	15,6	29,0	..	2,9
	7	16,1	5,7	..	10,4	21,0	..	..
	8	9,0	4,8	..	4,2	11,5	..	..
	9	10,1	7,2	..	2,9	10,7	..	..
	10	10,4	4,0	..	6,4	11,2	..	..
	11	11,0	5,9	..	5,1	15,6	..	..
	12	14,0	5,6	..	8,4	18,7	..	..
General Mean		14,73	5,13	..	9,60	19,60	..	2,91
		GOZAISHODAKE 1200m.						
IX 1888	4	20,3	16,6	18,45	4,7	75,5	100,2	16,1
	5	18,9	10,9	14,90	8,0	75,4	128,5	9,6
	6	19,7	9,6	14,65	10,1	76,5	125,6	8,0
	7	**21.7**	11,2	16,45	10,5	80,3	137,1	10,1
	8	18,0	13,7	15,85	4,3	68,8	92,1	10,9
	9	19,2	13,1	16,15	6,1	76,3	116,1	11,4
	10	18,5	13,7	16,10	4,8	66,8	94,5	13,6
	11	15,8	13,5	14,65	2,3	..	..	..
	12	18,3	13,4	15,85	4,9	66,5	100,0	12,0
	13	19,1	7,1	13,40	12,0	77,7	131,6	7,4
	14	15,5	8,3	16,90	7,2	71,3	120,9	7,7
	15	19,5	9,0	14,25	10,5	76,0	**137.3**	8,8
	16	17,3	11,0	14,15	6,3	68,1	129,6	11,0
	17	16,6	10,8	13,70	5,8	70,1	120,7	8,5
	18	18,0	8,4	13,20	9,6	74,4	126,3	6,7
	19	17,2	8,4	12,80	8,8	69,3	119,8	1,9
	20	16,2	14,0	15,10	2,2	61,1	85,5	10,6
	21	16,8	13,5	15,15	3,3	68,6	96,9	13,5
	22	17,6	13,0	15,30	4,6	72,1	128,2	13,4
	23	19,2	11,5	16,35	7,7	77,7	127,2	7,1
	24	18,4	8,1	13,25	10,3	75,9	142,8	6,5
	25	14,8	7,1	10,95	7,7	62,3	89,5	3,7
	26	17,6	9,7	13,65	7,9	72,8	114,2	7,7
	27	20,7	10,8	15,75	9,9	**82.1**	134,4	5,6
	28	18,5	10,6	14,55	7,9	78,5	122,8	5,8
	29	17,9	8,4	13,15	9,5	74,5	129,0	7,0
	30	17,1	7,0	12,05	10,1	75,9	132,1	5,6
X 1888	1	14,4	5,8	10,10	8,6	68,6	117,2	4,7
	2	10,6	3,7	7,15	6,9	58,2	114,5	2,9
	3	12,0	**3.3**	7,65	8,7	61,5	112,8	**98.5**
	Mean	17,5	10,2	13,85	7,3	71,6	118,2	8,6

TABLE VIIIa. EXTREME AIR TEMPERATURE AND RADIATION.

Month & Year	Day	Max. Temp.	Min. Temp.	Mean Temp.	Range	Max. S. R.	Max. S.R.V.	Min. T.R.
		KUROSAWA 832m.						
		°	°	°	°	°	°	°
VIII 1891	20	25,9	9,6	..	16,3	38,6	..	6,4
	21	19,9	13,3	..	6,6	24,2	..	12,0
	22	22,1	17,2	..	4,9	26,4	..	16,4
	23	20,7	15,8	..	4,9	22,8	..	18,7
	24	24,0	14,0	..	10,0	33,3	..	12,2
	25	26,4	14,6	..	11,8	40,3	..	12,7
	26	27,6	11,3	..	16,3	40,5	..	8,4
	27	28,8	8,4	..	20,4	42,8	..	5,3
	28	27,4	11,4	..	16,0	39,4	..	8,7
	29	28,8	12,7	..	16,1	41,0	..	9,4
	30	29,1	8,9	..	20,2	40,7	..	5,4
	31	29,1	9,9	..	19,2	40,8	..	7,0
	Mean	27,64	14,60	..	13,04	39,78	..	12,45
IX 1891	1	29,6	13,9	..	15,7	43,2	..	11,6
	2	30,2	16,0	..	14,2	46,5	..	13,8
	3	31,5	16,5	..	15,0	44,4	..	15,1
	4	29,5	16,5	..	13,0	42,8	..	13,8
	5	30,5	15,1	..	15,4	42,9	..	12,4
	6	30,1	13,9	..	16,2	43,3	..	11,4
	7	27,2	12,3	..	14,9	38,4	..	8,8
	8	25,0	16,2	..	8,8	31,4	..	14,8
	9	24,5	17,6	..	6,9	32,7	..	17,0
	10	27,4	18,0	..	9,4	38,7	..	16,5
	11	29,9	16,2	..	13,7	41,7	..	13,4
	12	30,6	14,6	..	16,0	41,0	..	12,4
General Mean		27,97	14,87	..	13,10	39,98	..	12,72
		YOKKAICHI 4m.						
IX 1888	4	**32.7**	23,3	28,00	9,4	**41.2**	**68.4**	22,2
	5	28,2	20,0	24,10	8,2	32,7	61,2	17,5
	6	28,9	16,1	22,50	12,8	34,9	60,6	13,0
	7	28,6	18,0	23,30	10,6	36,2	60,1	15,2
	8	25,1	19,3	22,20	5,8	27,7	35,8	17,0
	9	25,5	20,2	22,85	5,3	32,4	53,6	18,6
	10	23,0	19,0	21,00	4,0	23,9	31,8	18,8
	11	25,1	19,2	22,15	5,9	27,7	41,8	18,2
	12	28,6	21,4	25,00	7,2	37,2	66,2	20,0
	13	27,1	16,3	21,70	10,8	34,3	60,7	13,0
	14	25,9	15,4	20,65	10,5	30,0	61,8	11,9
	15	26,6	15,0	20,80	11,6	35,4	62,2	13,0
	16	22,6	18,8	20,70	3,8	28,2	41,6	15,5
	17	27,1	16,1	21,60	11,0	32,9	50,9	14,1
	18	27,5	15,4	21,45	12,1	35,4	61,3	11,5
	19	26,2	13,8	20,00	12,4	35,7	61,8	10,6
	20	23,6	20,9	22,25	2,7	28,2	53,0	19,8
	21	23,6	20,1	21,85	3,5	25,5	37,4	20,0
	22	23,4	20,0	21,70	3,4	31,0	61,8	19,0
	23	28,6	17,0	22,80	11,6	35,7	62,9	14,0
	24	26,7	14,2	20,45	12,5	33,8	62,9	10,9
	25	23,1	14,4	18,75	8,7	32,7	61,6	11,2
	26	25,8	16,9	21,35	8,9	34,3	60,4	15,0
	27	25,1	16,9	21,00	8,2	33,8	59,3	14,0
	28	26,8	12,4	19,60	14,4	33,8	57,4	9,3
	29	26,7	14,2	20,45	12,5	36,0	60,6	11,0
	30	24,7	13,6	19,15	11,1	27,7	59,7	9,8
X 1888	1	24,6	14,9	19,75	9,7	34,7	62,4	11,5
	2	21,5	11,7	16,60	9,8	26,7	49,6	8,0
	3	22,2	**8.3**	15,25	13,9	29,1	50,2	**4.0**
	Mean	25,8	16,8	21,30	9,0	32,5	56,2	14,3

TABLE VIII. EXTREME AIR TEMPERATURE.

Month & Year	Day	Max.	Min.	Mean	Range
		HIGASHI HOBEN 786^m.			
VIII 1889	1	26.7	20.6	23.65	6.1
	2	27.2	21.0	24.10	6.2
	3	27.6	22.2	24.90	5.4
	4	25.6	21.8	23.70	3.8
	5	28.7	22.1	25.40	6.6
	6	27.6	21.0	24.30	6.6
	7	27.4	20.5	23.95	6.9
	8	28.2	20.2	24.20	8.0
	9	28.9	20.3	24.60	8.6
	10	27.0	21.2	24.10	5.8
	11	29.1	20.7	24.90	8.4
	12	26.9	19.6	23.25	7.3
	13	28.2	20.0	24.10	8.2
	14	23.7	19.1	21.40	4.6
	15	22.1	18.7	20.40	3.4
	16	23.8	18.3	21.05	5.5
	17	25.9	20.0	22.95	5.9
	18	26.9	20.6	23.75	6.3
	19	23.1	19.2	21.15	3.9
	20	22.2	19.3	20.75	2.9
	21	27.4	19.5	23.45	7.9
	22	27.7	20.6	24.15	7.1
	23	25.7	21.0	23.35	4.7
	24	24.2	20.7	22.45	3.5
	25	22.3	20.5	21.40	1.8
	26	21.6	17.1	19.35	4.5
	27	20.6	16.3	18.45	4.3
	28	21.3	16.7	19.00	4.6
	29	20.4	15.2	17.80	5.2
	30	21.6	16.2	18.90	5.4
	31	19.7	17.1	18.40	2.6
	Mean	25.14	19.60	22.37	5.54
IX 1889	1	21.5	17.0	19.25	4.5
	2	20.5	16.8	18.65	3.7
	3	22.8	15.5	19.15	7.3
	4	24.2	18.2	21.20	6.0
	5	23.9	18.3	21.10	5.6
	6	20.7	16.5	18.60	4.2
	7	22.1	16.2	19.15	5.9
	8	22.9	16.3	19.60	6.6
	9	21.4	15.5	18.45	5.9
	10	21.6	14.1	17.85	7.5
	11	17.8	14.4	16.10	3.4
	12	17.9	14.1	16.00	3.8
	13	18.6	13.2	15.90	5.4
	14	18.9	13.1	16.00	5.8
	15	20.6	13.2	16.90	7.4
	16	18.4	12.6	15.50	5.8
	17	19.5	13.0	16.25	6.5
	18	22.1	14.5	18.30	7.6
	19	20.2	14.9	17.55	5.3
	20	19.2	15.3	17.25	3.9
	21	19.5	15.4	17.45	4.1
	22	20.9	14.0	17.45	6.9
	23	16.9	13.2	15.05	3.7
	24	18.4	13.0	15.70	5.4
	25	18.1	13.1	15.60	5.0
	26	20.1	12.2	16.15	7.9
	27	17.3	12.9	15.10	4.4

TABLE VIII^b. EXTREME AIR TEMPERATURE.

Month & Year	Day	Max.	Min.	Mean	Range
		YAMAGUCHI 35^m.			
VIII 1889	1	32.2	22.7	27.45	9.5
	2	32.3	23.7	28.00	8.6
	3	34.1	22.4	28.25	11.7
	4	31.1	20.9	26.00	10.2
	5	33.3	20.6	26.95	12.7
	6	33.0	22.6	27.80	10.4
	7	32.9	23.9	28.40	9.0
	8	33.8	22.3	28.05	11.5
	9	34.1	20.3	27.20	13.8
	10	33.6	21.4	27.50	12.2
	11	34.2	22.4	28.30	11.8
	12	34.0	20.9	27.45	13.1
	13	33.5	20.8	27.15	12.7
	14	30.5	22.3	26.40	8.2
	15	30.5	23.7	27.10	6.8
	16	30.6	22.9	26.75	7.7
	17	32.1	24.2	28.15	7.9
	18	33.3	21.5	27.40	11.8
	19	31.9	24.7	28.30	7.2
	20	29.6	23.8	26.70	5.8
	21	33.5	21.3	27.40	12.2
	22	33.0	22.5	27.75	10.5
	23	31.9	24.8	28.35	7.1
	24	30.2	26.2	28.20	4.0
	25	27.9	24.6	26.25	3.3
	26	27.7	22.8	25.25	4.9
	27	28.8	21.9	25.35	6.9
	28	29.0	21.6	25.30	7.4
	29	28.5	18.4	23.45	10.1
	30	28.2	20.2	24.20	8.0
	31	24.9	19.6	22.25	5.3
	Mean	31.43	22.32	26.87	9.11
IX 1889	1	28.0	20.7	24.35	7.3
	2	27.2	21.0	24.10	6.2
	3	29.9	18.5	24.20	11.4
	4	29.1	18.5	23.80	10.6
	5	29.4	21.2	25.30	8.2
	6	27.5	21.3	24.40	6.2
	7	29.3	18.5	23.90	10.8
	8	29.5	16.6	23.05	12.9
	9	28.6	17.3	22.95	11.3
	10	28.2	17.0	22.60	11.2
	11	24.7	19.8	22.25	4.9
	12	25.4	18.3	21.85	7.1
	13	26.3	13.8	20.05	12.5
	14	26.9	12.7	19.80	14.2
	15	25.4	12.2	18.80	13.2
	16	25.8	10.5	18.15	15.3
	17	26.8	12.6	19.70	14.2
	18	27.9	11.8	19.85	16.1
	19	26.3	13.5	19.90	12.8
	20	23.7	19.1	21.40	4.6
	21	24.3	18.9	21.60	5.4
	22	26.0	16.3	21.15	9.7
	23	22.3	15.3	18.80	7.0
	24	24.9	13.0	18.95	11.9
	25	23.0	12.5	17.75	10.5
	26	25.4	11.8	18.60	13.6
	27	20.7	17.8	19.25	2.9

TABLE VIII. EXTREME AIR TEMPERATURE.

Month & Year	Day	Max.	Min.	Mean	Range
HIGASHI HOBEN 736m.					
IX 1889	28	16,9°	13,1°	15,00°	3,8°
	29	15,7	11,3	13,50	4,4
	30	14,2	9,6	11,90	4,6
	Mean	19,76	14,35	17,05	5,41
X 1889	1	15,9	9,4	12,65	6,5
	2	17,4	10,1	13,75	7,3
	3	19,1	11,1	15,10	8,0
	4	21,1	12,2	16,65	8,9
	5	19,9	13,9	16,90	6,0
	6	16,8	12,5	14,65	4,3
	7	16,9	11,8	14,35	5,1
	8	17,4	11,3	14,35	6,1
	9	16,0	11,6	13,80	4,4
	10	18,4	12,1	15,25	6,3
	11	19,9	12,7	16,30	7,2
	12	20,4	10,3	15,35	10,1
	13	17,1	11,8	14,45	5,3
	14	17,0	9,7	13,35	7,3
	15	15,1	11,5	13,30	3,6
	16	15,0	11,3	13,15	3,7
	17	17,9	10,7	14,30	7,2
	18	18,5	11,8	15,15	6,7
	19	18,5	12,4	15,45	6,1
	20	16,0	11,4	13,70	4,6
	21	17,8	7,5	12,65	10,3
	22	9,2	6,0	7,60	3,2
	23	11,7	6,2	8,95	5,5
	24	14,9	7,6	11,25	7,3
	25	17,3	8,0	12,65	9,3
	26	15,6	8,6	12,10	7,0
	27	17,9	10,3	14,10	7,6
	28	13,9	12,0	12,95	1,9
	29	13,5	0,9	7,20	12,6
	30	8,6	0,4	4,50	8,2
	31	12,0	4,2	8,10	7,8
	Mean	16,34	9,72	13,03	6,63

TABLE VIII'. EXTREME AIR TEMPERATURE.

Month & Year	Day	Max.	Min.	Mean	Range
YAMAGUCHI 35m.					
IX 1889	28	24,0°	15,4°	19,70°	8,6°
	29	22,9	15,8	19,35	7,1
	30	20,9	12,3	16,60	8,6
	Mean	20,01	16,13	21,07	9,88
X 1889	1	22,4	7,4	14,90	15,0
	2	21,9	9,8	15,85	12,1
	3	25,1	11,5	18,30	13,6
	4	25,7	10,7	18,20	15,0
	5	25,0	10,3	17,65	14,7
	6	22,6	17,1	19,85	5,5
	7	24,3	12,0	18,15	12,3
	8	24,5	11,6	18,05	12,9
	9	23,1	13,5	18,30	9,6
	10	25,3	10,4	17,85	14,9
	11	24,3	8,8	16,55	15,5
	12	25,2	13,8	19,50	11,4
	13	24,3	9,2	16,75	15,1
	14	22,4	7,8	15,10	14,6
	15	20,9	14,9	17,90	6,0
	16	22,4	14,1	18,25	8,3
	17	23,7	9,4	16,55	14,3
	18	24,2	12,2	18,20	12,0
	19	22,3	10,2	16,25	12,1
	20	23,2	11,5	17,35	11,7
	21	22,5	13,1	17,80	9,4
	22	16,8	8,5	12,65	8,3
	23	18,5	4,0	11,25	14,5
	24	18,6	8,3	13,45	10,3
	25	20,5	5,1	12,80	15,4
	26	22,2	7,2	14,70	15,0
	27	23,3	10,2	16,75	13,1
	28	20,5	14,3	17,40	6,2
	29	17,2	3,7	10,45	13,5
	30	14,2	0,4	7,30	13,8
	31	16,2	4,5	10,35	11,7
	Mean	22,01	9,85	15,91	12,19

Month & Year	Day	ON SURFACE 2^h am	6^h am	10^h am	2^h pm	6^h pm	10^h pm	Mean	BELOW 3.0m 10^h am	10^h pm	Mean
						ONTAKE 3062m.					
VIII 1891	1	° ..	° ..	° 14.8	° 24.7	° 13.6	° 9.3	° ..	° ..	° 10.2	° ..
	2	8.9	9.3	11.6	10.2	8.8	7.9	9.45	9.1	9.6	9.35
	3	8.6	9.0	11.8	11.7	8.9	8.3	9.72	8.9	9.0	8.90
	4	7.8	9.2	13.3	13.9	10.2	8.5	10.62	8.8	9.3	9.05
	5	8.1	8.7	13.8	17.3	11.1	9.4	11.40	8.7	10.1	9.40
	6	9.2	8.8	17.4	16.8	13.3	9.2	12.45	9.7	11.1	10.40
	7	8.3	8.7	20.6	25.8	16.7	8.2	14.72	10.0	12.8	11.40
	8	4.4	6.1	18.2	25.1	13.9	7.8	12.58	10.6	12.1	11.35
	9	6.1	6.1	20.4	18.9	11.1	8.3	11.82	10.3	11.5	10.90
	10	8.2	8.7	13.0	14.6	10.6	8.8	10.65	10.2	10.8	10.50
	11	7.8	7.3	11.0	13.5	10.9	8.6	9.77	9.7	10.0	9.85
	12	6.8	5.2	17.3	24.5	12.3	8.7	12.47	10.5	10.8	10.65
	13	9.6	8.9	16.5	22.8	13.8	8.1	13.28	9.7	11.8	10.75
	14	6.6	8.2	20.7	24.9	15.6	10.4	14.40	10.1	12.8	11.45
	15	8.6	6.2	14.9	22.6	10.1	8.6	11.83	10.8	11.5	11.15
	16	6.3	5.8	13.6	12.2	9.8	8.8	9.42	9.6	9.0	9.30
	17	..	..	10.2	..	10.1	9.8	..	9.0	9.3	9.15
	18	9.1	9.2	10.2	10.0	6.1	4.4	8.17	9.2	8.8	9.00
	19	1.7	1.0	13.8	22.3	6.3	1.6	7.78	7.2	9.0	8.10
	20	0.6	0.6	13.3	23.3	7.8	2.6	8.03	6.2	8.0	7.10
	21	3.8	4.8	7.6	8.4	7.8	8.4	6.80	6.0	7.0	6.50
	22	8.5	6.2	10.6	10.9	7.7	7.8	8.62	7.0	7.8	7.40
	23	7.5	7.2	9.0	10.1	7.1	7.1	8.00	7.5	8.1	7.80
	24	2.8	3.7	11.8	15.6	9.6	5.1	8.10	6.7	8.0	7.35
	25	4.4	3.9	13.7	23.2	8.6	1.7	9.25	7.0	8.5	7.75
	26	0.5	99.2	19.9	19.1	9.9	1.5	8.20	6.2	8.0	7.10
	27	0.2	98.9	17.1	15.2	7.8	3.8	7.17	6.0	7.5	6.75
	28	1.9	1.3	16.1	15.0	7.9	2.2	7.40	6.3	7.3	6.80
	29	2.9	4.7	13.3	20.3	9.1	1.8	8.68	6.0	7.5	6.75
	30	0.1	0.4	24.9	27.4	11.1	4.1	11.33	5.7	8.0	6.85
	31	2.0	2.5	23.3	20.0	6.8	5.0	9.93	6.2	8.0	7.10
	Mean	5.60	5.54	15.13	17.77	9.99	6.42	10.07	8.29	9.43	8.86
IX 1891	1	3.7	5.9	29.8	25.1	13.4	8.4	12.88	7.0	9.2	8.10
	2	8.6	7.8	17.1	20.5	..	8.6	..	9.0	10.2	9.60
	3	7.6	7.0	23.4	19.6	14.4	7.7	13.28	9.3	11.0	10.15
	4	6.1	6.1	20.6	22.9	11.7	6.6	12.50	9.8	11.2	10.50
	5	5.1	4.1	15.9	20.4	13.9	5.7	12.52	9.5	11.2	10.35
	6	4.2	4.2	..	..	..	..	..	..	..	..
	7	..	..	..	..	..	..	..	..	..	..
	8	..	..	..	..	..	..	..	..	..	..
	9	..	..	..	..	..	..	..	..	..	..
	10	..	..	..	..	..	..	..	..	..	..
	11	..	..	..	..	..	..	..	..	..	..
	12	..	..	..	..	..	..	..	..	..	..
General Mean		5.60	5.54	15.74	18.02	10.40	6.50	10.40	8.15	9.31	8.74

Month & Year	Day	2h am	4h am	6h am	8h am	10h am	Noon	2h pm	4h pm	6h pm	8h pm	10h pm	Mid-night	Mean
ON SURFACE														
KUROSAWA 832m.														
VIII 1891		°	°	°	°	°	°	°	°	°	°	°	°	°
	1	..	..	..	..	..	..	..	..	..	..	..	..	..
	2	..	..	..	..	..	..	..	..	..	..	..	..	..
	3	..	..	..	..	..	..	..	..	..	..	..	..	..
	4	..	..	..	..	..	..	..	..	..	..	..	..	..
	5	..	..	..	..	..	..	..	..	24,2	21,7	21,1	20,0	..
	6	20,0	20,0	20,0	24,8	37,2	37,1	37,6	28,6	26,7	22,2	21,4	20,2	26.32
	7	20,2	19,6	20,7	29,6	38,9	33,6	31,3	27,7	25,2	22,3	19,7	18,6	25.62
	8	18,6	16,7	16,9	31,2	35,1	41,9	39,8	35,3	26,6	20,9	16,8	18,4	26.52
	9	18,1	17,5	17,5	29,6	39,1	44,7	34,7	38,8	27,4	22,2	19,2	19,1	27.49
	10	18,6	18,6	18,8	29,3	30,1	41,6	45,3	34,6	27,2	18,6	20,0	19,4	26.84
	11	19,3	17,5	18,3	30,2	33,7	31,7	36,6	34,1	26,9	21,7	17,3	18,6	25.49
	12	17,6	16,7	17,8	31,4	43,8	50,9	..	..	..	21,9	20,3	19,7	..
	13	19,2	17,8	18,6	28,0	43,9	47,1	38,9	37,9	27,3	21,8	19,1	17,8	28.12
	14	16,4	15,2	16,1	32,9	40,0	..	..		29,9	24,9	23,3	17,3	..
	15	21,6	21,4	21,7	38,9	..	..	38,9	..	29,7	21,6	19,7	17,6	..
	16	16,5	14,8	15,7	32,4	51,0	..	..	32,1	24,3	21,3	20,5	19,6	..
	17	19,1	19,7	19,6	20,3	21,9	25,9	29,1	25,7	22,5	21,2	20,3	19,9	22.08
	18	19,8	19,8	20,1	21,7	26,6	36,6	34,1	24,9	21,9	17,7	16,8	15,5	22.88
	19	14,9	13,4	14,7	26,3	37,1		..	..	20,3	16,2	14,6	12,9	..
	20	12,0	10,9	12,9	25,8	35,3	35,8	36,4	32,4	22,6	16,9	14,9	14,8	22.56
	21	15,2	15,1	16,1	16,2	19,8	19,0	20,1	20,7	19,6	18,9	18,6	18,4	18.14
	22	18,4	18,0	17,8	23,0	25,8	28,7	26,1	25,5	21,0	19,7	19,2	18,5	21.81
	23	18,1	17,3	18,3	20,9	24,2	24,4	23,1	22,7	20,5	18,6	17,4	16,3	20.15
	24	16,3	15,7	15,7	22,5	28,1	32,8	29,1	26,3	21,8	19,3	18,4	17,9	21.90
	25	16,5	16,1	17,6	27,1	36,2	36,2	36,6	28,1	23,1	18,0	16,9	16,3	24.06
	26	14,6	14,2	15,5	19,4	35,4	40,0	40,3	34,2	22,4	16,5	19,8	12,6	23.74
	27	10,9	10,0	9,8	25,9	35,9	42,1	41,2	32,4	23,2	19,7	16,7	15,0	23.57
	28	15,2	13,1	12,7	21,5	34,8	40,1	44,3	35,8	23,6	18,6	16,2	14,8	24.22
	29	15,5	15,4	15,7	20,8	34,6	45,8	46,3	35,2	24,5	18,6	15,8	13,7	25.16
	30	12,2	10,6	11,8	19,9	39,2	48,9	43,3	29,0	22,6	17,6	15,2	13,4	23.64
	31	11,9	11,3	11,5	36,7	42,2	46,7	45,8	36,8	24,2	19,0	18,0	18,6	26.89
	Mean	16.65	15.94	16.47	24.97	33.24	37.18	36.29	30.75	23.84	19.52	17.99	17.04	24.16
IX 1891	1	15,5	14,3	14,9	29,4	47,2	45,2	36,4	36,9	25,3	20,2	19,0	17,6	26.82
	2	17,2	16,7	18,3	28,8	40,7	..	34,8	28,3	22,4	19,4	18,8	18,8	..
	3	17,6	17,5	18,3	30,1	47,1	..	..	39,7	25,5	22,2	19,4	20,3	..
	4	19,7	18,1	18,2	28,7	39,0	46,3		..	26,1	20,2	17,8	16,4	..
	5	14,3	17,4	17,3	30,0	47,8	..	43,8	31,9	24,2	19,3	17,3	15,4	..
	6	15,8	17,3	16,5	25,3	47,7	45,7	..	35,2	23,8	19,0	18,5	15,4	..
	7	14,4	13,1	14,4	24,8	41,8	39,9	38,8	32,3	23,7	20,6	18,3	18,4	25.04
	8	18,0	17,7	17,3	26,0	32,4	35,3	31,8	27,2	21,8	19,7	18,8	18,8	23.73
	9	18,4	18,4	18,4	19,9	22,4	29,3	27,7	29,0	22,7	21,2	20,3	20,1	22.32
	10	20,0	19,3	19,6	28,2	37,2	33,8	33,6	30,7	24,3	22,3	22,1	19,2	25.86
	11	18,6	16,8	18,3	28,3	36,8	36,5	38,8	32,7	24,4	20,0	18,9	17,3	25.60
	12	17,6	16,3	15,3	27,1	39,1	40,9	39,2	32,9	23,9	20,9	21,1	21,0	26.27
General Mean		16.86	16.10	16.57	25.20	34.10	37.20	36.01	30.98	23.81	19.85	18.44	17.51	24.39

Plate I. View of Fuji from Lake of Yamanaka.

Plate II. View of Meteorological Station at Fuji.

Plate III. The Summit of Fuji.

Plate IV. View of Ontake from Kurosawa.

Plate V. View of Ontake from Nishinonohara.

Plate VI. The Summit of Ontake.

明治二十六年四月十九日印刷
明治二十六年四月二十日出版
中央氣象臺
印刷者　印　刷　局

www.ingramcontent.com/pod-product-compliance
Lightning Source LLC
Chambersburg PA
CBHW021955190326
41519CB00009B/1268
9783744790864